Lecture Notes in Computer Science 12988

More information about this subseries at https://link.springer.com/bookseries/7409

Jinpeng Wei · Liang-Jie Zhang (Eds.)

Big Data – BigData 2021

10th International Conference
Held as Part of the Services Conference Federation, SCF 2021
Virtual Event, December 10–14, 2021
Proceedings

 Springer

Editors
Jinpeng Wei 🆔
University of North Carolina
Charlotte, NC, USA

Liang-Jie Zhang 🆔
Kingdee International Software Group
Co., Ltd.
Shenzhen, China

ISSN 0302-9743   ISSN 1611-3349 (electronic)
Lecture Notes in Computer Science
ISBN 978-3-030-96281-4   ISBN 978-3-030-96282-1 (eBook)
https://doi.org/10.1007/978-3-030-96282-1

LNCS Sublibrary: SL3 – Information Systems and Applications, incl. Internet/Web, and HCI

This Springer imprint is published by the registered company Springer Nature Switzerland AG
The registered company address is: Gewerbestrasse 11, 6330 Cham, Switzerland

Preface

The International Congress on Big Data aims to provide an international forum that formally explores various business insights on all kinds of value-added "services". Big data is a key enabler of exploring business insights and economics of services.

BigData is a member of the Services Conference Federation (SCF). SCF 2021 comprised the following 10 collocated service-oriented sister conferences: the International Conference on Web Services (ICWS 2021), the International Conference on Cloud Computing (CLOUD 2021), the International Conference on Services Computing (SCC 2021), the International Conference on Big Data (BigData 2021), the International Conference on AI and Mobile Services (AIMS 2021), the World Congress on Services (SERVICES 2021), the International Conference on Internet of Things (ICIOT 2021), the International Conference on Cognitive Computing (ICCC 2021), the International Conference on Edge Computing (EDGE 2021), and the International Conference on Blockchain (ICBC 2021).

As the founding member of the Services Conference Federation (SCF), the First International Conference on Web Services (ICWS) was held in June 2003 in Las Vegas, USA. A sister event, the First International Conference on Web Services - Europe 2003 (ICWS-Europe 2003) was held in Germany in October of the same year. In 2004, ICWS-Europe was changed to the European Conference on Web Services (ECOWS), which was held in Erfurt, Germany. Sponsored by the Services Society and Springer, SCF 2018 and SCF 2019 were held successfully in Seattle and San Diego, USA, respectively. SCF 2020 was held successfully online and in Shenzhen, China. The 19th edition in the series, SCF 2021, was held during December 10–14, 2021.

This volume presents the accepted papers for BigData 2021, held as a fully virtual conference, during December 10–14, 2021. The major topics of BigData 2021 included, but were not limited to, big data architecture, big data modeling, big data as a service, big data for vertical industries (government, healthcare, etc.), big data analytics, big data toolkits, big data open platforms, economic analysis, big data for enterprise transformation, big data in business performance management, big data for business model innovations and analytics, big data in enterprise management models and practices, big data in government management models and practices, and big data in smart planet solutions.

We accepted eight papers, including six full papers and two short papers. Each paper was reviewed by at least three independent members of the BigData 2021 Program Committee. We are pleased to thank the authors whose submissions and participation made this conference possible. We also want to express our thanks to the Organizing Committee and Program Committee members for their dedication in helping to organize the conference and reviewing the submissions. We are grateful to all volunteers, authors, and conference participants for their contribution to the fast-growing worldwide services innovations community.

December 2021

Jinpeng Wei
Liang-Jie Zhang

Organization

BigData 2021 General Chair

Lakshmish Ramaswamy   University of Georgia, USA

BigData 2021 Program Chair

Jinping Wei   University of North Carolina at Charlotte, USA

Services Conference Federation (SCF 2021)

General Chairs

Wu Chou   Essenlix Corporation, USA
Calton Pu   Georgia Tech, USA
Dimitrios Georgakopoulos   Swinburne University of Technology, Australia

Program Chairs

Liang-Jie Zhang   Kingdee International Software Group Co., Ltd,
                                        China
Ali Arsanjani   Amazon Web Services, USA

Industry Track Chairs

Awel Dico   Etihad Airways, UAE
Rajesh Subramanyan   Amazon Web Services, USA
Siva Kantamneni   Deloitte Consulting, USA

CFO

Min Luo   Huawei, USA

Industry Exhibit and International Affairs Chair

Zhixiong Chen   Mercy College, USA

Operations Committee

Jing Zeng   Shenguodian Co., Ltd., China
Yishuang Ning   Tsinghua University, China
Sheng He   Tsinghua University, China

Steering Committee

Calton Pu (Co-chair)	Georgia Tech, USA
Liang-Jie Zhang (Co-chair)	Kingdee International Software Group Co., Ltd, China

BigData 2021 Program Committee

Muhammad Usama Javaid	nonE
Verena Kantere	University of Ottawa, Canada
Harald Kornmayer	DHBW Mannheim, Germany
Ugur Kursuncu	University of South Carolina, USA
Eugene Levner	Holon Institute of Technology, Israel
Yu Liang	University of Tennessee at Chattanooga, USA
Sagar Sharma	Wright State University, USA
Luiz Angelo Steffenel	Université de Reims Champagne-Ardenne, France
Sam Supakkul	NCR Corporation, USA
Nan Wang	Heilongjiang University, China
Wenbo Wang	Harrisburg University, USA

Conference Sponsor – Services Society

The Services Society (S2) is a non-profit professional organization that has been created to promote worldwide research and technical collaboration in services innovations among academia and industrial professionals. Its members are volunteers from industry and academia with common interests. S2 is registered in the USA as a "501(c) organization", which means that it is an American tax-exempt nonprofit organization. S2 collaborates with other professional organizations to sponsor or co-sponsor conferences and to promote an effective services curriculum in colleges and universities. S2 initiates and promotes a "Services University" program worldwide to bridge the gap between industrial needs and university instruction.

The services sector accounted for 79.5% of the GDP of the USA in 2016. Hong Kong has one of the world's most service-oriented economies, with the services sector accounting for more than 90% of GDP. As such, the Services Society has formed 10 Special Interest Groups (SIGs) to support technology and domain specific professional activities:

- Special Interest Group on Web Services (SIG-WS)
- Special Interest Group on Services Computing (SIG-SC)
- Special Interest Group on Services Industry (SIG-SI)
- Special Interest Group on Big Data (SIG-BD)
- Special Interest Group on Cloud Computing (SIG-CLOUD)
- Special Interest Group on Artificial Intelligence (SIG-AI)
- Special Interest Group on Edge Computing (SIG-EC)
- Special Interest Group on Cognitive Computing (SIG-CC)
- Special Interest Group on Blockchain (SIG-BC)
- Special Interest Group on Internet of Things (SIG-IOT)

About the Services Conference Federation (SCF)

As the founding member of the Services Conference Federation (SCF), the First International Conference on Web Services (ICWS) was held in June 2003 in Las Vegas, USA. A sister event, the First International Conference on Web Services - Europe 2003 (ICWS-Europe 2003) was held in Germany in October of the same year. In 2004, ICWS-Europe was changed to the European Conference on Web Services (ECOWS), which was held at Erfurt, Germany. SCF 2020 was held successfully. The 19th edition in the series, SCF 2021 was held virtually over the Internet during December 10–14, 2021.

In the past 18 years, the ICWS community has expanded from Web engineering innovations to scientific research for the whole services industry. The service delivery platforms have expanded to mobile platforms, the Internet of Things (IoT), cloud computing, and edge computing. The services ecosystem has gradually been enabled, value added, and intelligence embedded through enabling technologies such as big data, artificial intelligence, and cognitive computing. In the coming years, transactions with multiple parties involved will be transformed by blockchain.

Based on the technology trends and best practices in the field, SCF will continue serving as the conference umbrella's code name for all services-related conferences. SCF 2021 defined the future of the New ABCDE (AI, Blockchain, Cloud, big Data, Everything is connected), which enable IoT and support the "5G for Services Era". SCF 2021 featured 10 collocated conferences all centered around the topic of "services", each focusing on exploring different themes (e.g. web-based services, cloud-based services, big data-based services, services innovation lifecycle, AI-driven ubiquitous services, blockchain driven trust service-ecosystems, industry-specific services and applications, and emerging service-oriented technologies). The SCF 2021 members were as follows:

1. The 2021 International Conference on Web Services (ICWS 2021, http://icws.org/) was the flagship event for web-based services, featuring web services modeling, development, publishing, discovery, composition, testing, adaptation, and delivery, as well as the latest API standards.
2. The 2021 International Conference on Cloud Computing (CLOUD 2021, http://thecloudcomputing.org/) was the flagship event for modeling, developing, publishing, monitoring, managing, and delivering XaaS (everything as a service) in the context of various types of cloud environments.
3. The 2021 International Conference on Big Data (BigData 2021, http://bigdatacongress.org/) focused on the scientific and engineering innovations of big data.
4. The 2021 International Conference on Services Computing (SCC 2021, http://thescc.org/) was the flagship event for the services innovation lifecycle, including enterprise modeling, business consulting, solution creation, services orchestration, services optimization, services management, services marketing, and business process integration and management.
5. The 2021 International Conference on AI and Mobile Services (AIMS 2021, http://ai1000.org/) addressed the science and technology of artificial intelligence and the

development, publication, discovery, orchestration, invocation, testing, delivery, and certification of AI-enabled services and mobile applications.

6. The 2021 World Congress on Services (SERVICES 2021, http://servicescongress. org/) put its focus on emerging service-oriented technologies and industry-specific services and solutions.

7. The 2021 International Conference on Cognitive Computing (ICCC 2021, http:// thecognitivecomputing.org/) put its focus on Sensing Intelligence (SI) as a Service (SIaaS), which makes a system listen, speak, see, smell, taste, understand, interact, and/or walk, in the context of scientific research and engineering solutions.

8. The 2021 International Conference on Internet of Things (ICIOT 2021, http:// iciot.org/) addressed the creation of IoT technologies and the development of IOT services.

9. The 2021 International Conference on Edge Computing (EDGE 2021, http://theedg ecomputing.org/) put its focus on the state of the art and practice of edge computing including, but not limited to, localized resource sharing, connections with the cloud, and 5G devices and applications.

10. The 2021 International Conference on Blockchain (ICBC 2021, http://blockchai n1000.org/) concentrated on blockchain-based services and enabling technologies.

Some of the highlights of SCF 2021 were as follows:

- Bigger Platform: The 10 collocated conferences (SCF 2021) got sponsorship from the Services Society which is the world-leading not-for-profits organization (501 c(3)) dedicated to serving more than 30,000 services computing researchers and practitioners worldwide. A bigger platform means bigger opportunities for all volunteers, authors, and participants. In addition, Springer provided sponsorship for best paper awards and other professional activities. All 10 conference proceedings of SCF 2021 will be published by Springer and indexed in the ISI Conference Proceedings Citation Index (included in the Web of Science), the Engineering Index EI (Compendex and Inspec databases), DBLP, Google Scholar, IO-Port, MathSciNet, Scopus, and ZBlMath.

- Brighter Future: While celebrating the 2021 version of ICWS, SCF 2021 highlighted the Fourth International Conference on Blockchain (ICBC 2021) to build the fundamental infrastructure for enabling secure and trusted services ecosystems. It will also lead our community members to create their own brighter future.

- Better Model: SCF 2021 continued to leverage the invented Conference Blockchain Model (CBM) to innovate the organizing practices for all 10 collocated conferences.

Contents

Research Track

A Combination of Resampling and Ensemble Method for Text Classification on Imbalanced Data

Haijun Feng, Wen Qin, Huijing Wang, Yi Li, and Guangwu Hu(✉)

Shenzhen Institute of Information Technology, Shenzhen 518172, China
hugw@sziit.edu.cn

Abstract. One of the major factor which can affect the accuracy of text classification is the imbalanced dataset. In order to find the suitable method to handle this issue, six different ensemble methods are used to train models on imbalanced dataset. The result shows that, without resampling the dataset, Stacking algorithm performs better than other ensemble method, it can increase the recall of the minority class by 19.3%. Meanwhile, ensemble algorithms combined with resampling methods are used to train the model. Results show that, ensemble algorithms combined with undersampling method (RUS) can improve the predictive ability of models on minority class, but it reduces the accuracy of models on other majority classes because of feature dropping; while voting algorithm combined with oversampling method (SmoteTomek) can improve the recall of the minority class by 40.4%, without decreasing the accuracy of models on other majority classes. Afterall, in training a text classification model with multi-class imbalanced datasets, Voting algorithm combined with SmoteTomek can be a preference.

Keywords: Imbalanced data · Voting · Resampling · Text classification · Ensemble

1 Introduction

Text classification is one of the major applications of nature language process, it's main goal is to label the text automatically by computer, which needs dataset and suitable algorithm to train the model and predict results. But the precision of the model depends on the quality of dataset and algorithm. One factor which could reduce the precision of the model is that the dataset is imbalanced. It means in the dataset, some classes have a large number of samples which is called major classes, while some classes have very few samples which is called minor classes. Models trained by this dataset have a very poor prediction ability on minor classes. In order to solve the imbalanced issue of dataset, lots of researches are carried out. Improvements are mainly based on two directions, the dataset level and the algorithm level [1–3].

© Springer Nature Switzerland AG 2022
J. Wei and L.-J. Zhang (Eds.): BigData 2021, LNCS 12988, pp. 3–16, 2022.
https://doi.org/10.1007/978-3-030-96282-1_1

On the dataset level, the main strategy is to use resampling methods, oversampling and undersampling methods will be introduced to resample the data to get a balanced dataset. Popular resampling methods are Smote [4], RandomOverSampler, RandomUnderSampler, SmoteTomek [5] and so on. Krawczyk [6] proposes a multiclass radial-based oversampling (MC-RBO) method, This method uses potential functions to generate artificial instances which can handle multiclass imbalanced data. Douzas [7] proposes an oversampling method based on k-means clustering and SMOTE, this method avoids the generation of noise when generating new samples and it overcomes imbalances between and within classes effectively. Liu [8] develops a fuzzy-based information decomposition (FID) method, which rebalances the training data by creating synthetic samples for the minority class with weighting and recovery processes. This method performs well on lots of well-known datasets.

On the algorithm level, the main idea is adjusting the algorithms to improve the accuracy of models, such as cost-sensitive learning, ensemble method. Feng [9] proposes an ensemble margin based algorithm, in this algorithm, ensemble learning and undersampling are combined to construct higher quality balanced datasets. This method shows good effectiveness in class imbalance learning. Zhao [10] proposes a decomposition-based approach, they use an unsupervised learning algorithm to learn the hidden structure of the majority class firstly, and then transform the classification problem into several classification sub-problems. The performance of this method is good on various datasets. Cao [11] proposes a novel method which is called Feature Inverse Mapping based Cost-sensitive Stacking learning (IMCStacking). In IMCStacking, the cost-sensitive Logistic Regression is integrated as the final classifier to regard different costs to majority and minority samples. This method can handle the imbalanced classification problems. Bader-El-Den [12] proposes a novel and effective ensemble-based method for handling the class imbalanced issue. In the method, they proposed a biased random forest algorithm which identifies the critical areas by the nearest neighbor algorithm. The proposed algorithm is very effective in handling the class imbalanced issue. Yuan [13] proposes a method which employs regularization, it accommodates multi-class data sets and determines the error bound automatically. In this method, the classifier is penalized when it misclassifies examples that were correctly classified in the previous learning phase. This strategy can improve the maximum accuracy to 24.7%.

Combining the resampling and ensemble method together to train models could be a way to overcome the imbalanced issue. In this work, we carry out our experiment in two aspects. First we use six ensemble methods (Bagging [14], Random Forest [15], AdaBoost [16], GradientBoosting [17], Voting, Stacking [18]) to train the model and analyze the effect on improving the accuracy of models. Then we use RandomUnderSampler and SmoteTomek [5] algorithms to resample the dataset. After this, we combine this two aspects together to train the models, which is the biggest innovation in our work. We also discuss how resampling algorithms and ensemble methods improve the accuracy of model trained with imbalance dataset.

The rest of the paper is organized as follows: Sect. 2 introduces our methods and strategies to handle imbalanced issue. Section 3 introduces our experiment, discusses and compares the effect of different combined methods. Section 4 gives the conclusions and recommends the best combined method in dealing with imbalanced datasets.

2 Methods and Experiments

2.1 Methods

The framework of the experiment is showed as Fig. 1. We first clean the corpus, segment the Chinese words and remove stop words. After this preprocessing work, we transform the Chinese words into bag of words and TF-IDF [19]. Then we carry out parallel experiments by two roads. In road 1, six different ensemble methods (Bagging [14], Random Forest [15], AdaBoost [16], GradientBoosting [17], Voting, Stacking [18]) are used to train the model and we evaluate the classifier; in road 2, the dataset is resampled by different algorithm and then trained with ensemble methods such as EasyEnsemble [20], RUSBoost [21]. We discuss the effect of these different methods in overcoming the imbalanced issue and how they increase the recognition rate of models on minor classes.

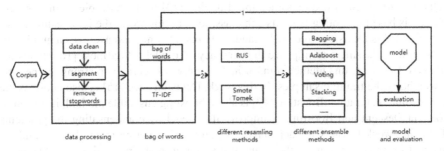

Fig. 1. Workflow of the method

2.2 Experimental Details

We use python language and Jupyter notebook to carry out the work, SciKit-learn [22] and imbalanced-learn [23] libraries are called, all jobs are running on Baidu AI studio.

The dataset in this experiment is collected from customers' remarks of retailers online. Five categories are chose, including mobile phones, fruits, water heaters, clothes and computers. The total number of samples is 26856. The profile of different classes is showed in Table 1, from which we can get that the major class fruit has 9992 samples, the minor class water heaters has 573 samples, the ratio is 17.4:1, this dataset is imbalanced. In training the model, we pick 80% of the dataset out randomly as a training set to train the model, the left 20% are used as a test set for testing. Then we use the model to classify the text.

In the experiment, we use jieba tool to segment the Chinese words and remove the stop words based on the Chinese stop words list. Next, we call scikit-learn [22] to build the bag of words and transform words into Term Frequency-Inverse Document Frequency [19] (TF-IDF). We remove high frequency words which occur in 80% of samples, and we also remove low frequency words which occur less than 3 times. Finally, 8934 feature words remain after the preprocessing.

Table 1. Profile of 5 shopping categories

	Items	Number
1	Fruits	9992
2	Clothes	9985
3	Computers	3983
4	Mobile phones	2323
5	Water heaters	573

Six different ensemble methods are used to train the models in this experiment, Decision Tree [15] method is treated as a baseline as it is a common machine learning method wildly used in text classification. Bagging [14] is an ensemble method whose strategy is based on voting from its sub classifiers, in this work, the base_estimator of Bagging [14] is Decision Tree [15]. Random Forest [15] method averages a number of decision tree classifiers to improve the accuracy of models. AdaBoost [16] stands for Adaptive Boosting, this method aims at improving weak learners to strong learners by re-weighting the training samples. In this work, Decision Tree [15] is also chose as the base_estimator. GradientBoosting [17] is also a boosting ensemble method, it combines Gradient Descent and Boosting, it improve the model by compensating the gradients of existing weak learners in each stage. The strategy of voting method is to combine different machine learning methods and use voting to make predictions. In this work, Random Forest [15], Naive Byes [19], Logistic Regression [24], Neural Network [25] are combined to predict class labels by soft voting. Stacking method is also an ensemble method which uses the output of different classifiers as input of a final classifier to get a more accurate model. It takes advantage of each classifier's strength. In this work, Random Forest [15], Naive Byes [19], LinearSVC method are used to train the base classifiers, and Logistic Regression [24] method are followed to train the final classifier with above output as input.

RandomUnderSampler and SmoteTomek [5] algorithms are used to resample the original training dataset, RandomUnderSampler (RUS) is a random undersampling methods, which removes samples in the major classes randomly and produce a new training dataset. SmoteTomek [5] is a combination method of oversampling method Smote [4] and undersampling method TomekLinks [26], it uses Smote method to generate new minor samples to the training dataset and then clean the TomekLinks [26] in the new dataset. The profile of the new dataset after resampling is showed in Table 2, the two resampling method can both balance the dataset, but RUS method drops lots of feature words.

Table 2. Training set profile of 5 shopping categories after resampling

	Fruits	Clothes	Computers	Mobile phones	Water heaters
Original training set	7978	8026	3187	1835	458
RandomUnderSampler	458	458	458	458	458
SmoteTomek	7948	7946	8025	8022	8023

3 Results and Discussions

3.1 Evaluation Metrics

Lots of metrics are used to evaluate a model, such as overall accuracy, precision, recall, f1score. In this text classification, five categories are included, which is a multi-class issue. In this situation, we use confusion matrix to calculate recall of each class, and also we calculate the overall accuracy (acc), precision, recall and f1score. We evaluate the model by these factors, which are defined as below respectively.

$$\Pr ecision = \frac{TP}{TP + FP} \tag{1}$$

$$Recall = \frac{TP}{TP + FN} \tag{2}$$

$$f1score = 2 * \frac{\Pr ecision * Recall}{\Pr ecision + Recall} \tag{3}$$

$$acc = \frac{TP + TN}{TP + TN + FP + FN} \tag{4}$$

In the formulas, TP stands for true positives, TN for true negatives, FP for false positives and FN for false negatives.

3.2 Results of Different Ensemble Methods

In this text classification of five categories, Decision Tree [15] method is used to train the model as a baseline, then six ensemble algorithms are carried out to train the model with original dataset respectively. We analyzed the effect of each ensemble method, detailed results are showed in Table 3. Common machine learning method "Decision Tree [15]" behaves bad in handling imbalanced issue, the model trained by Decision Tree has a bad recognition ratio on minor class "water heaters", with a low recall 0.539 and the overall accuracy is only 0.879. Ensemble method "Bagging [14]" and "Random Forest [15]" can improve the overall accuracy of models, but for the minor class "water heaters", the recall is 0.530 and 0.496 respectively, without any improvement compared to "Decision Tree" method. Adaboost [16] method performs bad, it cannot improve the recall of model on minor class "water heaters", and what's worse, it reduces the overall accuracy of the model. GradientBoosting [17] method contributes little to the improvement of prediction ability of model on minor class, but it increase the overall accuracy of model to a certain extent. Voting method has similar effects, increasing the overall accuracy but cannot improve the recall of model on minor class. Stacking [18] method behaves best, it not only improve the prediction ability of model on minor class, but also increase the overall accuracy. The recall of minor class "water heaters" is improved from 0.539 to 0.643, the increasing degree is 19.3%, and meanwhile, the overall accuracy rises from 0.879 to 0.951, an increasing degree of 8.2%. Precision and F1score reflect the same results, so in training a model with an imbalanced dataset, Stacking [18] algorithm can enhance the prediction ability of models better than other ensemble methods. But still the accuracy of model on minor class is not high enough to make prediction, it needs to be improved.

Table 3. Metrics of model trained by different ensemble algorithms with original dataset

Class	Method	Recall	Precision	F1score	Acc.
Mobile phones	Decision Tree	0.768	0.847	0.806	0.879
	Random Forest	0.891	0.940	0.915	0.922
	Bagging	0.777	0.875	0.823	0.897
	AdaBoost	0.621	0.781	0.692	0.830
	Gradient Boosting	0.826	0.953	0.885	0.909
	Voting	0.893	0.958	0.925	0.935
	Stacking	0.945	0.958	0.951	0.951
Fruits	Decision Tree	0.905	0.907	0.906	0.879
	Random Forest	0.929	0.940	0.934	0.922
	Bagging	0.919	0.928	0.923	0.897
	AdaBoost	0.865	0.872	0.868	0.830
	Gradient Boosting	0.910	0.950	0.929	0.909
	Voting	0.947	0.943	0.945	0.935
	Stacking	0.952	0.959	0.956	0.951
Water heaters	Decision Tree	0.539	0.633	0.582	0.879
	Random Forest	0.496	0.826	0.620	0.922
	Bagging	0.530	0.693	0.601	0.897
	AdaBoost	0.322	0.649	0.430	0.830
	Gradient Boosting	0.522	0.789	0.628	0.909
	Voting	0.504	0.921	0.652	0.935
	Stacking	0.643	0.860	0.736	0.951

(*continued*)

Table 3. (*continued*)

Class	Method	Recall	Precision	F1score	Acc.
Clothes	Decision Tree	0.915	0.880	0.897	0.879
	Random Forest	0.954	0.889	0.920	0.922
	Bagging	0.941	0.881	0.910	0.897
	AdaBoost	0.875	0.865	0.870	0.830
	Gradient Boosting	0.965	0.853	0.905	0.909
	Voting	0.956	0.917	0.936	0.935
	Stacking	0.961	0.939	0.950	0.951
Computers	Decision Tree	0.842	0.851	0.846	0.879
	Random Forest	0.908	0.968	0.937	0.922
	Bagging	0.857	0.892	0.874	0.897
	AdaBoost	0.830	0.699	0.759	0.830
	Gradient Boosting	0.876	0.956	0.914	0.909
	Voting	0.943	0.952	0.948	0.935
	Stacking	0.970	0.963	0.966	0.951

We can get the effect of each ensemble method intuitively from Fig. 2. In handling imbalanced dataset, AdaBoost [16] performs not very well compared to other methods, the recall of model on minor class water heaters is smallest by this method, which means it cannot improve the prediction ability of model on minor class and also this method is time-consuming. Stacking [18] method gets the best performance among all those ensemble methods, in the stacking method, weak learners are combined to build a strong learner, this strategy can make the learners complement each other and reduce mistakes made by model prediction. The other ensemble methods can improve the overall accuracy of models to a certain extent, but they help little in distinguishing the minor class water heaters, which has the fewest samples. In the CPU running process, AdaBoost [16] and GradientBoosting [17] methods are both time-consuming.

3.3 Results of Resampling Algorithms Combined with Ensemble Methods

In order to find a better method to overcome the imbalanced problem, we combine resampling algorithms with ensemble methods to train the models in this work. We first use four algorithms (BalancedRandomForest [27], BalancedBagging [28], EasyEnsemble [20], RUSBoost [21]) to train the model with original dataset, this four algorithms are all based on undersampling methods. BalancedRandomForest [27] and BalancedBagging [28] algorithms are based on RandomForest [15] and Bagging [14] algorithms respectively, they drop samples randomly to balance the training set during training. EasyEnsemble [20] algorithm is also dropping samples randomly to balance the AdaBoost [16] learners

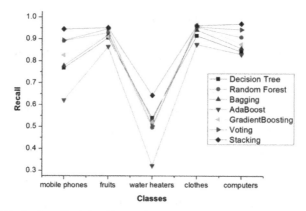

Fig. 2. Recall of classifiers trained by different ensemble method

during training and then assemble the AdaBoost [16] learners together. RUSBoost [21] algorithm is a combination of RandomUnderSampler and AdaBoost [16] algorithms, this method drops samples randomly each boosting step to alleviate the imbalanced issue. Results of different methods is showed in Table 4, the recognition rate of models on each class is balanced, the prediction of models will not prior to major classes any more. But lots of feature samples are dropped during training, the overall accuracy, precision and f1score are all reduced compared to the baseline of Decision Tree [15] method. Meanwhile, RandomUnderSampler are also used to resample the dataset, the profile of resampled dataset is showed in Table 2. Each class has only 458 samples after under sampling. Then GradientBoosting [17], Voting, Stacking [18] methods are used to train the models respectively with the under resampled dataset. The results is showed in Table 4, the effect is similar to the above four ensemble method, although this method can alleviate the preference of models on major classes, but the overall accuracy, precision and f1score are all reduced because of feature samples dropping.

Table 4. Metrics of models trained by different ensemble methods with dataset under resampled

Class	Method	Recall	Precision	F1score	Acc.
Mobile phones	Balanced RandomForest	0.881	0.879	0.880	0.860
	Balanced Bagging	0.865	0.714	0.782	0.843
	EasyEnsemble	0.867	0.801	0.833	0.830
	RUSBoost	0.809	0.809	0.809	0.798
	Gradient Boosting	0.855	0.832	0.843	0.821
	Voting	0.932	0.890	0.911	0.878

(continued)

Table 4. (*continued*)

Class	Method	Recall	Precision	F1score	Acc.
	Stacking	0.949	0.869	0.907	0.891
Fruits	Balanced RandomForest	0.882	0.950	0.915	0.860
	Balanced Bagging	0.855	0.958	0.903	0.843
	EasyEnsemble	0.850	0.952	0.898	0.830
	RUSBoost	0.820	0.935	0.874	0.798
	Gradient Boosting	0.838	0.956	0.893	0.821
	Voting	0.889	0.956	0.921	0.878
	Stacking	0.894	0.958	0.925	0.891
Water heaters	Balanced RandomForest	0.904	0.187	0.310	0.860
	Balanced Bagging	0.817	0.214	0.339	0.843
	EasyEnsemble	0.870	0.168	0.281	0.830
	RUSBoost	0.861	0.133	0.231	0.798
	Gradient Boosting	0.870	0.139	0.240	0.821
	Voting	0.930	0.232	0.371	0.878
	Stacking	0.896	0.294	0.443	0.891
Clothes	Balanced RandomForest	0.826	0.934	0.877	0.860
	Balanced Bagging	0.831	0.909	0.868	0.843
	EasyEnsemble	0.793	0.927	0.854	0.830
	RUSBoost	0.765	0.904	0.829	0.798
	Gradient Boosting	0.784	0.929	0.850	0.821
	Voting	0.836	0.938	0.884	0.878
	Stacking	0.857	0.930	0.892	0.891
Computers	Balanced RandomForest	0.866	0.952	0.907	0.860
	Balanced Bagging	0.832	0.877	0.854	0.843
	EasyEnsemble	0.843	0.869	0.856	0.830
	RUSBoost	0.809	0.898	0.851	0.798
	Gradient Boosting	0.839	0.910	0.873	0.821
	Voting	0.916	0.935	0.925	0.878
	Stacking	0.935	0.924	0.929	0.891

From Fig. 3(a), we can get the results intuitively, the recall of models which are trained by 7 methods with dataset under resampled is close to each other, and the value is among 0.8–0.9. These strategies can improve the prediction ability of models on minor classes at the expense of the overall accuracy of models.

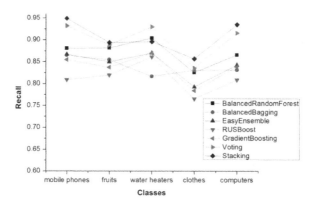

(a) Under Sampling

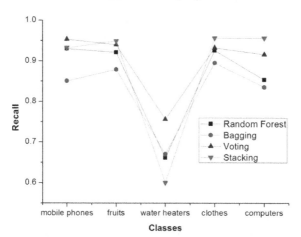

(b) SmoteTomek Sampling

Fig. 3. Recall of models trained by different ensemble methods combined with resampling algorithms

Other than under sampling method, we also use SmoteTomek [5] algorithm to resample the dataset, SmoteTomek [5] algorithm is a combination method of under and over sampling. The profile of resampled dataset is showed in Table 2, which is balanced. And then, Bagging [14], Random Forest [15], Voting, Stacking [18] methods are carried out to train the model respectively. The recall of each model is showed in Table 5 and Fig. 3(b). After resampling, models trained by Bagging [14], Random Forest [15], Voting methods all have an improvement in predicting minor class like water heaters without descending the overall accuracy. Especially for voting method, it has the biggest improvement, the recall is improved to 0.757, up to 40.4% compared to the baseline Decision Tree [15] method. While model trained by Stacking [18] method don't have an obvious improvement in predicting minor class after resampling. From the above discussions, we can get that Voting method combined with SmoteTomek [5] can overcome the imbalanced issue of dataset and have an better performance on multi-class text classification with imbalanced dataset.

Table 5. Metrics of models trained by different ensemble methods with dataset resampled by SmoteTomek algorithm

Class	Method	Recall	Precision	F1score	Acc.
Mobile phones	Random Forest	0.930	0.783	0.850	0.908
	Bagging	0.850	0.741	0.792	0.871
	Voting	0.953	0.871	0.910	0.931
	Stacking	0.932	0.930	0.931	0.944
Fruits	Random Forest	0.921	0.942	0.931	0.908
	Bagging	0.879	0.939	0.908	0.871
	Voting	0.939	0.956	0.947	0.931
	Stacking	0.949	0.958	0.954	0.944
Water heaters	Random Forest	0.661	0.524	0.585	0.908
	Bagging	0.670	0.377	0.483	0.871
	Voting	0.757	0.551	0.637	0.931
	Stacking	0.600	0.750	0.667	0.944
Clothes	Random Forest	0.926	0.914	0.920	0.908
	Bagging	0.895	0.888	0.891	0.871
	Voting	0.933	0.944	0.938	0.931
	Stacking	0.956	0.937	0.946	0.944
Computers	Random Forest	0.853	0.980	0.912	0.908
	Bagging	0.835	0.890	0.862	0.871
	Voting	0.916	0.953	0.934	0.931
	Stacking	0.956	0.955	0.955	0.944

Detailed prediction of models can be displayed clearly by confusion matrix. Figure 4 is the confusion matrix of model trained by Voting method combined with SmoteTomek [5] resampling algorithm. From this figure, we can get that the model can distinguish minor class like water heaters, most samples can be predicted correctly, and only a few samples can be predicted as fruits or clothes. Meanwhile, all the other classes can be predicted precisely, the overall accuracy of model is high.

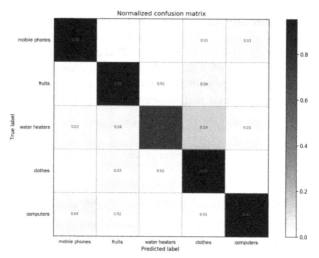

Fig. 4. Confusion matrix of model trained by Voting method combined with SmoteTomek resampling algorithm

Figure 5 shows the recall of models trained by Voting method, combined with RUS and SmoteTomek methods. With no resampling involved, model trained by Voting method has a bad prediction on the minor class "water heaters". When RUS resampling method is introduced, Recall of models has an obvious improvement on the minor class "water heaters", but this undersampling method decrease the predictive ability of model on other classes. When oversampling method SmoteTomek is combined, Recall of minor class "water heaters" is also improved, and Recall of other classes has almost no reduction. So, in handling multi-class imbalanced datasets, Voting method combined with SmoteTomek algorithm is a better strategy.

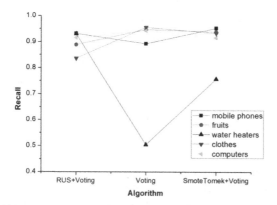

Fig. 5. Recall of models trained by Voting method combined with resampling algorithms

4 Conclusions

Imbalanced dataset causes inaccuracy in text classification, which makes the prediction of model prior to major class. In order to solve this issue, six different ensemble methods (Bagging, Random Forest, AdaBoost, GradientBoosting, Voting, Stacking) are carried out to train the model. Among all those ensemble algorithm, Stacking method has the best performance; it can improve the prediction ability of model on minor class with a degree of 19.3%. but the improvement is still not high enough to make precise predictions on minor classes.

Meanwhile, resampling algorithms and ensemble methods are combined together to train the models. Ensemble methods combined with undersampling algorithms can improve the preference of models on minor classes, but it will also reduce the overall accuracy of models because of feature data dropping. Voting method combined with SmoteTomek algorithm can overcome the imbalanced issue of dataset and improve the prediction ability of models on minor class "water heaters" by 40.4%. Plus, this strategy won't reduce the overall accuracy. So In training a model with imbalanced dataset, Voting method combined with SmoteTomek resampling algorithm can be a preference.

Acknowledgement. This work is supported by Ph.D. research project from Shenzhen Institute of Information Technology (NO: SZIIT2021KJ012) and the College-Enterprise Collaboration Project of Shenzhen Institute of Information Technology (11400-2021-010201-010199).

References

1. Wasikowski, M., Chen, X.W.: Combating the small sample class imbalance problem using feature selection. IEEE Trans. Knowl. Data Eng. **22**(10), 1388–1400 (2010)
2. Suh, S., Lee, H., Lukowicz, P., et al.: CEGAN: classification enhancement generative adversarial networks for unraveling data imbalance problems. Neural Netw. **133**, 69–86 (2021)
3. Kumari, C., Abulaish, M., Subbarao, N.: Using SMOTE to deal with class-imbalance problem in bioactivity data to predict mTOR inhibitors. SN Comput. Sci. **1**(3) (2020)
4. Chawla, N.V., Bowyer, K.W., Hall, L.O., et al.: SMOTE: synthetic minority over-sampling technique. J. Artif. Intell. Res. **16**(1), 321–357 (2002)
5. Batista Gustavo, E.A.P., Bazzan Ana, L.C., Monard, M.C.: Balancing training data for automated annotation of keywords: a case study. In: II Brazilian Workshop on Bioinformatics, pp. 10–18 (2008)
6. Krawczyk, B., Koziarski, M., Wozniak, M.: Radial-based oversampling for multiclass imbalanced data classification. IEEE Trans. Neural Netw. Learn. Syst. **PP**(99), 1–14 (2019)
7. Douzas, G., Bacao, F., Last, F.: Improving imbalanced learning through a heuristic oversampling method based on k-means and SMOTE. Inf. Sci. **465**, 1–20 (2018)
8. Liu, S., Zhang, J., Xiang, Y., et al.: Fuzzy-based information decomposition for incomplete and imbalanced data learning. IEEE Trans. Fuzzy Syst. **25**(6), 1476–1490 (2017)
9. Feng, W., Huang, W., Ren, J.: Class imbalance ensemble learning based on the margin theory. Appl. Sci. **8**(5), 815–843 (2018)
10. Zhao, Y., Shrivastava, A.K., Tsui, K.L.: Imbalanced classification by learning hidden data structure. IIE Trans. **48**(7), 614–628 (2016)
11. Cao, C., Wang, Z.: IMCStacking: cost-sensitive stacking learning with feature inverse mapping for imbalanced problems. Knowl.-Based Syst. **150**, 27–37 (2018)

12. Bader-El-Den, M., Teitei, E., Perry, T.: Biased random forest for dealing with the class imbalance problem. IEEE Trans. Neural Netw. Learn. Syst. **30**(7), 2163–2172 (2019)
13. Yuan, X., Xie, L., Abouelenien, M.: A regularized ensemble framework of deep learning for cancer detection from multi-class, imbalanced training data. Pattern Recogn. **77**, 160–172 (2018)
14. Breiman, L.: Bagging predictors. Mach. Learn. **24**(2), 123–140 (1996)
15. Breiman, L.: Random forests. Mach. Learn. **45**(1), 5–32 (2001)
16. Zhu, J.: Multi-class AdaBoost. Stat. Interf. **2**, 349–360 (2009)
17. Friedman, J.H.: Greedy function approximation: a gradient boosting machine. Ann. Stat. **29**(5), 1189–1232 (2001)
18. Wolpert, D.H.: Stacked generalization. Neural Netw. **5**(2), 241–259 (1992)
19. Manning, C.D., Raghavan, P., Schütze, H.: Introduction to Information Retrieval. Cambridge University Press, Cambridge (2008)
20. Liu, X.Y., Wu, J., Zhou, Z.H.: Exploratory undersampling for class-imbalance learning. IEEE Trans. Syst. Man Cybern. Part B **39**(2), 539–550 (2009)
21. Seiffert, C., Khoshgoftaar, T.M., Van Hulse, J., et al.: RUSBoost: a hybrid approach to alleviating class imbalance. IEEE Trans. Syst. Man Cybern. Part A Syst. Hum. **40**(1), 185–197 (2010)
22. Swami, A., Jain, R.: Scikit-learn: machine learning in Python. J. Mach. Learn. Res. **12**(10), 2825–2830 (2011)
23. Lemaitre, G., Nogueira, F., Aridas, C.K.: Imbalanced-learn: a Python toolbox to tackle the curse of imbalanced datasets in machine learning. J. Mach. Learn. Res. **18**(17), 1–5 (2017)
24. Yu, H.F., Huang, F.L., Lin, C.J.: Dual coordinate descent methods for logistic regression and maximum entropy models. Mach. Learn. **85**(1–2), 41–75 (2011)
25. Glorot, X., Bengio, Y.: Understanding the difficulty of training deep feedforward neural networks. J. Mach. Learn. Res. **9**, 249–256 (2010)
26. Tomek, I.: Two modifications of CNN. IEEE Trans. Syst. Man Cybern. **SMC-6**(11), 769–772 (1976)
27. Chen, C., Breiman, L.: Using random forest to learn imbalanced data. Univ. Calif. Berkeley **110**, 1–12 (2004)
28. Breiman, L.: Pasting small votes for classification in large databases and on-line. Mach. Learn. **36**(1–2), 85–103 (1999)

Validating Business Problem Hypotheses: A Goal-Oriented and Machine Learning-Based Approach

Robert Ahn[1(✉)], Sam Supakkul[2], Liping Zhao[3], Kirthy Kolluri[1], Tom Hill[4], and Lawrence Chung[1]

[1] University of Texas at Dallas, Richardson, TX, USA
{robert.sungsoo.ahn,kirthy.kolluri,chung}@utdallas.edu
[2] NCR Corporation, Atlanta, GA, USA
sam.supakkul@ncr.com
[3] University of Manchester, Manchester, UK
liping.zhao@manchester.ac.uk
[4] Fellows Consulting Group, Dallas, TX, USA

Abstract. Validating an elicited business problem hindering a business goal is often more important than finding solutions. For example, validating the impact of a client's account balance toward an unpaid loan would be critical as a bank can take some actions to mitigate the problem. However, business organizations face difficulties confirming whether some business events are against a business goal. Some challenges to validate a problem are discovering testable factors, preparing relevant data to validate, and analyzing relationships between the business events. This paper proposes a goal-oriented and Machine Learning(ML)-based framework, Gomphy, using a problem hypothesis for validating business problems. We present an ontology and a process, an entity modeling method for a problem hypothesis to find testable factors, a data preparation method to build an ML dataset, and an evaluation method to detect relationships among the business events and goals. To see the strength and weaknesses of our framework, we have validated banking events behind an unpaid loan in one bank as an empirical study. We feel that at least the proposed approach helps validate business events against a goal, providing some insights about the validated problem.

1 Introduction

The assertion that *"A problem unstated is a problem unsolved"* expresses the importance of eliciting business needs and problems [1]. Understanding and validating a business problem likely to hinder a business goal is often more critical than developing solutions. Validating a business problem helps define system boundaries in the early phase of requirements engineering [2]. If the correct problems are validated first, a business can save precious time and cost to deal with erroneous problems [3].

© Springer Nature Switzerland AG 2022
J. Wei and L.-J. Zhang (Eds.): BigData 2021, LNCS 12988, pp. 17–33, 2022.
https://doi.org/10.1007/978-3-030-96282-1_2

However, business organizations face difficulties confirming whether an elicited business event causes or impacts other high-level problems [4,5]. Specifically, some challenges might be identifying testable factors for the elicited problem, constructing a dataset to test, and determining whether the identified problem has some relationships and how many degrees towards the high-level problem and a business goal. Developing an information system with unconfirmed problems frequently leads to a system that is not useful enough to achieve business goals, costing valuable business resources [6,7].

Drawing on our previous work Metis [8], we present *GOMPHY*, a *G*oal-*O*riented and *M*achine learning-based framework using a *P*roblem *HY*pothesis, to help validate business problems [9,10]. This paper proposes four main technical contributions: 1. An ontology for modeling and validating a business problem hypothesis is described. 2. An entity modeling method for a problem hypothesis is presented to help identify an entity, attributes, constraints, and relationships for a problem hypothesis in the source dataset. 3. A data preparation method is described, mapping a problem hypothesis entity to a database entity and features, extracting and transforming a dataset. 4. An evaluation method is elaborated to detect positive or negative contributions among business problems and goals using Machine Learning (ML) and ML Explainability techniques.

This paper applies the proposed Gomphy framework to explore hypothesized business events behind an unpaid loan problem in one bank and validate the problem hypotheses as an empirical study. Figure 1 shows a high-level context diagram for the overdue loan problem. We use the PKDD'99 Financial database [11] to represent data that the bank may have managed.

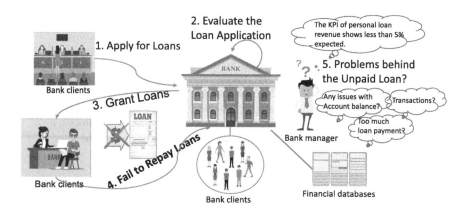

Fig. 1. Unpaid loan problem in a bank

The rest of this paper is structured as follows. Section 2 presents the Gomphy framework, and Sect. 3 illustrates the Gomphy process in detail with an unpaid loan problem. Next, Sect. 4 describes three experiments performed, and Sect. 5 discusses related work, observations, and limitations. Finally, Sect. 6 summarizes the paper and future work.

2 The Gomphy Framework

The Gomphy framework, aiming to help validate business problems, consists of a domain-independent ontology, a series of steps based on goal orientation (GO) and Machine Learning (ML).

2.1 The Gomphy Ontology

The ontology consists of categories of essential modeling concepts, relationships on modeling concepts, and constraints on the concepts and relationships, as shown in Fig. 2, where boxes and arrows represent the concepts and relationships.

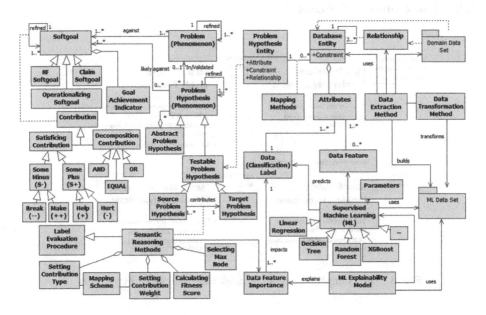

Fig. 2. The Gomphy ontology for a validating problem hypothesis

Some types of essential Gomphy concepts are introduced. A *(Soft-)Goal* is defined as a goal that may not have a clear-cut criterion and a *(Soft-)Problem* as a phenomenon against a Goal. A *Problem Hypothesis* is a hypothesis that we believe a phenomenon is against a Goal. There are two kinds of Problem Hypothesis, an *Abstract Problem Hypothesis* and a *Testable Problem Hypothesis*. An Abstract Problem Hypothesis is conceptual, whereas a Testable Problem Hypothesis is measurable. A Testable Problem Hypothesis may be further refined, forming a *Source Problem Hypothesis* and a *Target Problem Hypothesis*.

A *Problem Hypothesis Entity*, capturing a Testable Problem Hypothesis, is mapped to relevant *Database Entity*, *Attributes*, *Constraints*, and *Relationships* in a source data model. The selected Attributes are used to build an ML dataset consisting of *Data Features* and a *Classification Label*.

The Contribution relationships among Goals, Problems, and Problem Hypotheses are categorized into Decomposition types, such as *AND, OR, EQUAL*, or Satisficing types, such as *Make, Help, Hurt, Break, Some-Plus, Some-Minus* adopted from the NFR Framework [12]. The relationships between Problem Hypotheses and Problems are either *Validated* or *Invalidated*.

One crucial constraint about a problem hypothesis includes time-order among a target and a source problem hypothesis, where a source problem hypothesis must have occurred before the target problem hypothesis. Other constraints are a positive contribution from a source problem hypothesis to a target problem hypothesis, and the contribution should be reasonably sensible [13].

2.2 The Gomphy Process

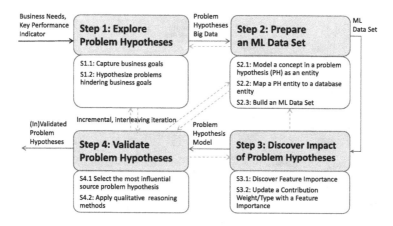

Fig. 3. The Gomphy process for validating a problem hypothesis

The Gomphy process, shown in Fig. 3, is intended to help guide steps for validating a problem hypothesis, providing traceability among a goal, a problem, a dataset, and ML. The process consists of four steps but should be understood as iterative, interleaving, and incremental in ML projects. The sub-steps of each step are described in detail in the following Sect. 3.

3 The Gomphy in Action

We suppose a hypothetical bank, the Case bank providing client services, such as opening accounts, offering loans, and issuing credit cards. The bank has experienced an unpaid loan problem. Some clients failed to make recurring payments when due. However, it did not know what specific clients' banking behaviors were behind this issue. Since this is a hypothetical example, we used the PKDD'99 Financial database to represent data the bank may have managed [11].

PKDD'99 Financial Database: The database contains records about banking services, such as Account (4,500 records), Transaction (1,053,620), Loan (682), Payment Order (6,471), and Credit cards (892). Six hundred six loans were paid off within the contract period, and seventy-six were not among the loan records. Figure 4 shows the schema of the Financial database.

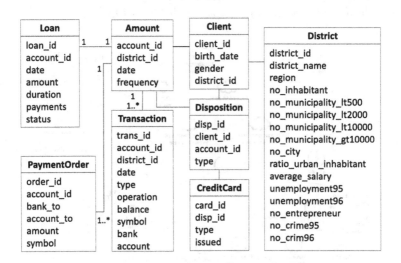

Fig. 4. The schema of the Financial database

3.1 Step 1: Explore the Case Bank's Problem Hypotheses

Requirements engineers begin Step 1, understanding and modeling the Case bank's goals. Potential problems hindering the goals are then hypothesized.

Step 1.1 Capture the Case Bank's Goals. After understanding the bank domain, one of the bank's goals, *Maximize revenue*$_{NFsoftgoal}$[1] is modeled as an NF (Non-Functional) softgoal to achieve at the top level, which is AND-decomposed and operationalized by *Increase loan revenue*$_{OPsoftgoal}$ and *Increase fee revenue*$_{OPsoftgoal}$ as operationalizing softgoals, as shown in Fig. 5. The former is further AND-decomposed to more specific softgoals of *Increase personal loan revenue*$_{OPsoftgoal}$ and *Increase business loan revenue*$_{OPsoftgoal}$.

During an interview, the bank staff indicated that the personal loan revenue of this quarter is less than 5% for the Key Performance Indicator (KPI) they intended to achieve due to some clients' unpaid loans. So, the bank wanted to know which specific banking events of a client contribute to the outstanding loan. However, the bank staff had difficulties with how to do that.

---

[1] The Gomphy concept is expressed in the notation from [14].

Step 1.2: Hypothesize Problems Hindering the Case Bank's Goal. We modeled that a client's *Unpaid loan$_{OPsoftproblem}$ Breaks*(−−) the *Increase personal loan revenue$_{OPsoftgoal}$*. After understanding the loan process and analysis of the Financial database, we explored potential clients' banking behaviors against the unpaid loan. We hypothesized that a client's *Loan$_{AbstractPH}$*, *Account Balance$_{AbstractPH}$*, and *Transaction$_{AbstractPH}$* might somewhat positively contribute to the *Unpaid loan$_{OPsoftproblem}$*, as shown in Fig. 5.

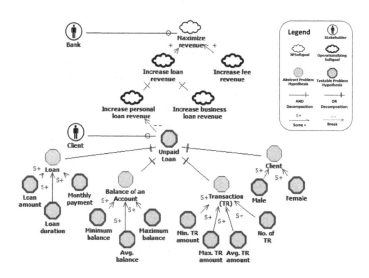

Fig. 5. Hypothesizing problems for an unpaid loan

An abstract problem hypothesis is further decomposed into a testable problem hypothesis that usually contains nominal, ordinal, interval, or ratio factors. For example, the *Balance of an Account$_{AbstractPH}$* may be divided into the more specific *Minimum balance of an Account$_{TestablePH}$*, *Average balance of an Account$_{TestablePH}$*, and *Maximum balance of an Account$_{TestablePH}$* for the client's loan duration using an OR-decomposition method.

Based on the goal and problem hypothesis graph above, we can express one of the problem hypotheses in a conditional statement. Let PH1 be the problem hypothesis *The minimum balance of an Account somewhat positively contributes to an unpaid loan for the loan duration$_{PH}$*. Then, we can consider the *Minimum balance of an Account$_{SourcePH}$* as a source problem hypothesis (or an independent variable), *Somewhat positively contributes$_{PHcontribution}$* as a contribution relationship, and an *Unpaid loan for the loan duration$_{TargetPH}$* as a target problem hypothesis (or a dependent variable).

$$\textit{Minimum balance of an Account}_{SourcePH}$$
$$\xrightarrow{\textit{Some-plus}_{PHcontribution}} \textit{Unpaid loan for the loan duration}_{TargetPH} \tag{1}$$

3.2   Step 2: Prepare an ML Dataset

In this step, we model a concept in the testable problem hypothesis as a problem hypothesis entity, maps the attributes of a problem hypothesis entity to the attributes of a database entity in the database or a domain dataset, and constructs an ML dataset based on the identified data attributes.

Step 2.1: Model a Concept in a Problem Hypothesis as an Entity. The concept in a testable problem hypothesis is modeled as an entity using the entity-relationship model [15]. The entity has attributes, constraints, and relationships. An *attribute* is a property of an entity having measurable value. A *constraint* is a condition restricting the value or state of a problem hypothesis. A *relationship* shows other entities associated with this entity.

For example, the *Minimum balance of an Account$_{SourcePH}$* in PH1 is modeled as a problem hypothesis entity of *Account$_{PHE}$*, having an attribute of *balance$_{PHEattribute}$*, a constraint of a *minimum balance$_{PHEconstraint}$*, and a relationship of a *Loan$_{PHErelationship}$*, as shown in Fig. 6.

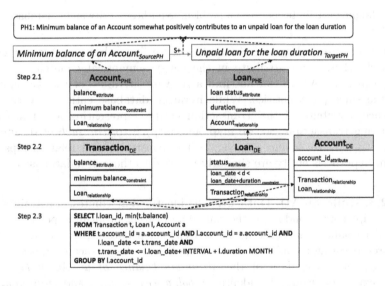

Fig. 6. Preparing an ML dataset for validating a problem hypothesis

Step 2.2: Map a Problem Hypothesis Entity to a Database Entity. The attribute in the problem hypothesis entity may manually be mapped to attributes in the database entity with tool support in Fig. 7. The tool first reads the database schema and shows the concerned database entity and attributes. We then select a database entity and check whether attributes in the entity are similar to the attributes of the problem hypothesis entity.

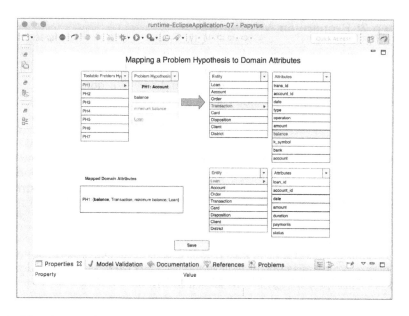

Fig. 7. Mapping attributes between a problem hypothesis entity and a database entity

For $balance_{PHEattribute}$ of $Account_{PHE}$, we first select the Account entity and check whether an attribute in the entity semantically matches the $balance_{PHEattribute}$. As we can not find a relevant attribute of the Account, we check the subsequent entities. While iterating database entities, we can find a 'balance' attribute of the Transaction entity, representing a balance after the banking transaction. So, we map $Account_{PHE}$ to $Transaction_{DE}$ and $balance_{PHEattribute}$ to $balance_{DEattribute}$. The constraint and relationships of a problem hypothesis entity are similarly mapped to those of a database entity.

Step 2.3: Build an ML Dataset. The identified attributes, constraints, and relationships corresponding to the source and target problem hypothesis entity are used to build a database query and extract a dataset, as shown in Fig. 6. Data preprocessing techniques are then applied to the extracted dataset.

For example, the data of the $Minimum\ balance\ of\ an\ Account_{SourcePH}$ can be extracted using the identified $balance_{DEattribute}$, and $minimum\ balance_{DEconstraint}$ in $Transaction_{DE}$. SQL group function, min() may be used to select $minimum\ balance_{DEconstraint}$. Also, to apply the relationship $Loan_{DErelationship}$, we need to identify a primary key and a foreign key relationship between $Loan_{DE}$ entity and $Transaction_{DE}$, which lead to identifying $Account_{DE}$ entity. The $loan\ duration_{DEconstraint}$ of $Loan_{DE}$ is also applied, as shown in Fig. 6.

We then tentatively store the resulting dataset for each testable problem hypothesis and integrate it into an ML dataset. Next, we may need to transform some feature values using preprocessing techniques, including scaling feature value using a normalization method and converting categorical data to a numeric value, such as using a one-hot encoding method, in our example on the transaction type, mode, symbol features. We may also fill in some missing values, replacing the null value with an average value and others [16].

3.3   Step 3: Discover the Impact of Problem Hypotheses Using ML

The impact of banking events towards the unpaid loan encoded as data features and a target label is uncovered using Supervised ML models and ML Explainability model, decoding hidden feature patterns in the dataset [17,18].

Step 3.1: Discover Feature Importance. To decode the relationships among banking events and an unpaid loan, four ML models, such as Logistic Regression (LR), Decision Tree (DT), Random Forest (RF), and eXtreme Gradient Boosting (XGBoost) were built with the dataset. The ML models then predicted the loan instances as 'Paid or Unpaid Loan.' The accuracy of each ML model was 0.92 (LR), 0.95 (DT), 0.973 (RF), and 0.977 (XGBoost), as shown in Fig. 13. The more accurate an ML model, the more confidence we can use feature importance value to validate a problem hypothesis.

Next, we utilized the SHAP (Shapley Additive exPlanations) model to get an intuitive and consistent feature value [19]. The XGBoost model was given as input to the SHAP model. To analyze the feature importance for prediction results, we first collected predicted instances of unpaid loans. Figure 8 shows the SHAP value of some important features for one case, where we can notice that the minimum balance, the minimum amount transaction, the average balance, and the household remittance somewhat positively impact the unpaid loan. The wider the width, the higher the impact. After that, we summed up the feature values of all the unpaid loans to detect the feature impact of all unpaid loans.

Fig. 8. Feature importance for one unpaid loaner

Step 3.2: Update a Contribution Weight and Type with Feature Importance. The collected feature importance value ($I_{source,target}$) can be considered a contribution weight from a source to a target problem hypothesis. The contribution weight and type of each leaf-level problem hypothesis are updated based on the detected feature importance value using Formula 2 and 3.

$$weight(PH_{source}, PH_{target}) = I_{source,target} \qquad (2)$$

$$ctr\_type\big(I_{source,target}\big) = \begin{cases} S+ & \text{if } I_{source,target} \geq 0 \\ S- & \text{if } I_{source,target} < 0 \end{cases} \qquad (3)$$

For example, the Contribution weight and type of the leaf node, the minimum balance, are updated with the detected value 15.32 and $S+$ in Fig. 9. Similarly, the contribution weight and type of other leaf nodes are updated accordingly.

Next, to know the direct and indirect impact of leaf-level problem hypotheses towards a high-level problem in the problem hypothesis model, we first calculate the fitness score of a source problem hypothesis using Formula 4.

$$score(PH_s) = \left(\sum_{t=1}^{\#targets} weight(PH_t) \times weight(PH_s, PH_t) \right) \quad (4)$$

We assume the weight of each problem hypothesis is 0.2, adopting a weight-based quantitative selection pattern [20]. For example, the fitness score of the *Minimum balance of an Account*$_{SourcePH}$ is calculated as $(0.2 * 15.32 =)$ 3.064.

3.4   Step 4: Validate Problem Hypotheses

This step selects the most critical problem hypothesis as a validated one among many alternative hypotheses and evaluates the impact of the validated problem on other high-level problems, as shown in Fig. 9.

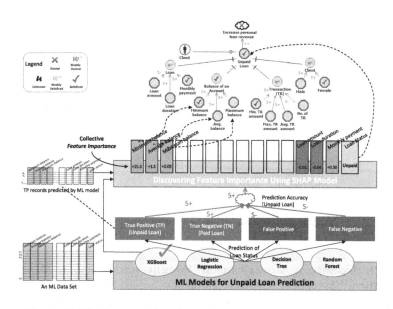

Fig. 9. Validating a problem hypothesis using feature importance

Step 4.1: Select the Most Influential Source Problem Hypothesis.
Among alternative problem hypotheses contributing to a target problem in the problem hypothesis model, we select a problem hypothesis having the highest fitness score in the leaf nodes. Banking staff may give a qualitative priority for

some problem hypotheses, depending on some schemes, such as 'normal', 'critical', or 'very critical'. Here, we assume a 'normal' priority for all the problem hypotheses.

For example, the *Minimum balance of an Account$_{SourcePH}$* in the problem hypothesis model was selected by Formula 5 as it has the highest fitness score among the leaf problem hypotheses under *Unpaid loan$_{OPsoftproblem}$*.

$$selection(PH_{target}) = max\Big(score(PH_{source})\Big)_{s=1}^{\#sources} \tag{5}$$

The selected problem hypothesis is considered a validated problem hypothesis by Formula 6, as it is most likely to be the cause for the target problem hypothesis [12]. It means the *Minimum balance of an Account$_{SourcePH}$* is likely to be the most important cause of the *Balance of an Account$_{AbstractPH}$*.

$$validated(selection(PH_i)) \rightarrow validated(PH_i) \tag{6}$$

Step 4.2: Apply Qualitative Reasoning Methods to Reason the Validation Impact Towards a High-Level Problem. Once the most likely problem hypothesis is validated, as shown by 'check mark' in Fig. 9, qualitative reasoning, e.g., the label propagation procedure [9], is carried out to determine the validated problem's impact upward a problem.

If the *Minimum balance of an Account$_{SourcePH}$* and *Somewhat positively contribute to$_{PHcontribution}$* are satisficed, then the *Balance of an Account$_{AbstractPH}$* is satisficed. The reasoning propagation shows that the *Balance of an Account$_{SourcePH}$* somewhat positively contributes to the *Unpaid loan$_{OPsoftproblem}$*, which *Breaks* the *Increase personal loan revenue$_{OPsoftgoal}$* in Fig. 9.

4   Experimental Results

We performed three experiments to see the strength and the weakness of Gomphy. While experiments 1 and 2 were performed without Gomphy, experiment 3 was conducted with the Gomphy framework.

4.1   Experiment 1

In this experiment, we treated all the features in the Financial database as potential events causing unpaid loans. We prepared the ML dataset by selecting all the data features in the database, except the table identifiers, where the features were considered potential problems and loan status as a target problem. The prepared ML dataset included 72 features with some transformation methods and 449,736 records based on the Transaction id. The large records are due to the *join* operation among Account, Transaction, and Payment Order tables.

As some ML algorithms such as Gradient Boosting Tree provide feature importance, we analyzed whether the provided essential features could be possible banking events leading to the unpaid loan. Figure 10(a) shows some important features predicted by the XGBoost model. However, it was not easy to get some ideas about whether the loan granted year and the credit card type, 'classic,' has some relationships towards the unpaid loan.

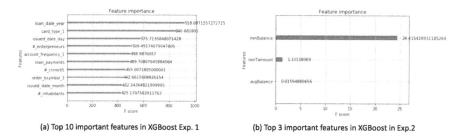

(a) Top 10 important features in XGBoost Exp. 1   (b) Top 3 important features in XGBoost in Exp.2

Fig. 10. Top important features in experiment 1 and 2

One critical issue of this approach is that the ML models, e.g., XGBoost, showed different prediction results for the same loan instance. For example, different transaction records, having the same Loan ID 233, showed different loan prediction results (i.e., paid and unpaid), which confused in identifying a banking event leading to the unpaid loan.

4.2   Experiment 2

In this experiment 2, the ML dataset was prepared based on the loan ID, unlike experiment 1, to understand the primary features produced by ML models. For preparing the loan-based dataset, we used SQL group functions, such as Sum, Min, and Avg, to select records for the one-to-many relationships between Account and Transactions. The final dataset contained 682 records, including 72 features. Four ML models were built to predict the loan instances. Figure 10(b) shows some important features for the XGBoost model.

Among the three important features, minimum balance, minimum transaction amount, and average balance, it could be possible that the minimum balance could cause an unpaid loan. However, it was confusing whether the minimum amount of transaction could lead to unpaid loan events. Other banking events related to the minimum amount of Transactions seemed to be needed to get a deep understanding of this issue.

A critical issue of this approach is that the prepared dataset did not consider the boundary of the records within the loan duration, which may give incorrect predictions and show a lack of rationale for identifying critical root causes of the unpaid loan. For example, when the loan duration of loan ID 1 is two years from 1993, the dataset included records of 1996 and 1997.

4.3   Experiment 3

In this experiment 3, we applied the Gomphy framework to validate clients' banking events towards the unpaid loan. The banking events were hypothesized

as four groups, including Loan, Account, Transaction, and Client. The hypothesis is further analyzed into testable problem hypotheses, as shown in Fig. 9.

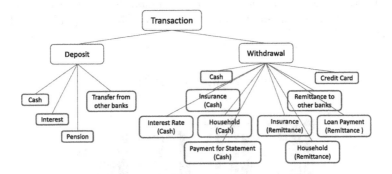

Fig. 11. Deposit and withdrawal classification in Transaction

While preparing a dataset, we could understand that the *balance* of Transaction depends on transaction type (deposit or withdrawal), operation (mode of a transaction), and symbol (characterization of the transaction) features, as shown in Fig. 11. So, we hypothesized events related to the operations and symbols as possible causes to the change of the balance, which were added to the set of a problem hypothesis. Otherwise, the category features would be hot-encoded in a usual ML approach, like in experiments 1 and 2.

Based on the modeled problem hypotheses with banking domain understanding and database analysis, six hundred eighty-two records with 25 features were prepared. Four ML models were run then to predict whether each loan could be paid off or not. Next, the ML Explainability model was applied with separately collected unpaid loan cases to understand better the impact of the features.

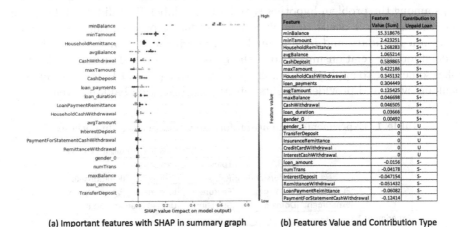

(a) Important features with SHAP in summary graph   (b) Features Value and Contribution Type

Fig. 12. Important features and contribution type in experiment 3

Figure 12, produced by the SHAP model, shows some important features for the unpaid loan, including the *minimum balance*, the *minimum transaction amount, remittance withdrawal for household cost*, and others. We could also understand that the *minimum transaction amount* is related to the *sanction interest* if the balance of an Account is negative after we performed further analysis.

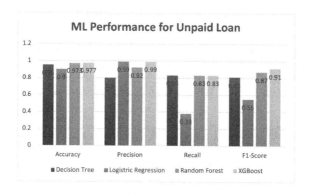

Fig. 13. ML models' performance comparison in experiment 3

The accuracy, precision, and F1-Score of the ML models in Experiment 3 are shown in Fig. 13. XGBoost showed slightly better accuracy than Random Forest. So, we selected XGBoost as the ML model for getting feature importance.

The trade-off analysis for the experiments is shown in Table 1. Experiment 1 is easier to perform the validation work assuming all the features as problem hypotheses. However, its results are difficult to understand, even giving different predictions for the same loan case. Experiment 2 shows more sensible results than experiment 1 but is still challenging to understand the rationale and the relationships among the problem hypotheses and a target label. To validate and explain potential events, it needs to apply some systematic process, data constraints, e.g., data boundary and feature analysis. Experiment 3 provides sensible and understandable relationships among the banking events and an unpaid loan. It takes some time to apply the Gomphy process but helps identify the most critical banking event and provides insights into the hypothesized banking events.

Table 1. Experiments comparison for validating problem hypotheses

Relative streangth: - < + < ++

	Easy to Experiment	Understanding of Banking Domain	Loan-Related Feature Selection	Feature Explanation towards Unpaid Loan	Relationship b/w Problem and Goal
Experiment 1	++	-	-	-	-
Experiment 2	+	+	-	+	-
Experiment 3	+	++	+	++	++

5   Discussion and Related Work

Problem analysis and validation have been studied to understand real-world problems in two major areas: Requirements Engineering and Machine Learning. The distinctive of our approach is to use a concept of a problem hypothesis to refute or confirm potential business problems using ML.

In Requirements Engineering, a Fishbone diagram has been used to identify possible causes for a problem or an effect [21]. This technique helps enumerate potential causes for a problem. However, the lack of a clear relationship between a cause and an effect, e.g., logical connectives, such as 'AND,' or 'OR', makes problem validation difficult. Fault Tree Analysis (FTA) is a top-down, deductive analysis that visually depicts a failure path or failure chain [22]. FTA provides Boolean logic operators. When linked in a chain, these statements form a logic diagram of failure. However, FTA does not provide relationship direction and degrees, such as positive, negative, full, and partial, making it challenging to validate business problems using ML. *(Soft-)Problem Inter-dependency Graph (PIG)* uses a (Soft−)problem concept to represent a stakeholder problem against stakeholder goals, where a problem is refined into sub-problems [23]. However, PIG lacks a mechanism to connect sub-problems to data features to test. While the Fishbone diagram, FTA, and PIG provide a sound high-level model, they need validation mechanisms for confirming the causes behind business problems.

In the area of Machine Learning, some ML algorithms, such as Linear Regression and Decision Trees, provide feature importance value concerning their predictions. When ML models predict a numerical value in the regression model or a target label in the classification, relative feature importance scores are calculated for the features in the dataset [3]. Explainable machine learning models also provide feature importance [24]. LIME(Local Interpretable Model-agnostic Explanations) explains individual predictions, but there is some instability of the explanations, which may hurt validating business problems [25]. SHAP (SHapley Additive exPlanations) outputs feature value that helps to understand business problems. However, SHAP may take a long computational time [19].

Although we could utilize feature importance in ML algorithms to get insights about business problems, one issue is identifying essential factors to test. Some features or attributes, among many features, in the dataset, might be redundant, irrelevant, or less critical to business problems. Dimensionality reduction techniques, such as Principal Component Analysis (PCA), and Formal Concept Analysis (FCA), are often used to find essential features for a target label [26,27]. However, the data features selected by the dimensionality reduction techniques often makes it difficult to understand transparent relationships between the features and high-level business problems in the context of goals [28]. Gomphy provides traceability from goals to problems and data features, bridging the gaps in a complimentary manner.

While some banking events, such as pension deposit at the end of a month, regularly occur, others, such as household payment with cash, randomly. Our work deals with the time-series nature of both banking events, but not in a strict sense. In-depth analysis and validation of the time-series banking events may provide other insights about potential causes of the unpaid loan.

Limitations. This paper has some limitations. 1) Correlation among problem hypotheses and goals could be utilized to understand the business events better, but the correlation analysis was not explored yet. 2) The mapping process between a problem hypothesis entity and a database entity is partially supported with a prototype mapping tool, although the tool needs more work to be more effective. 3) After a potential problem, e.g., a minimum balance, is validated, bank staff may take tentative actions. For example, the bank may waive fees on missed loan payments or offer affected clients options to defer loan payments for a finite period. The more long-term and effective solutions need to be explored and validated using a solution hypothesis to mitigate the validated problem.

6   Conclusion and Future Work

This paper has presented the Gomphy framework to validate business problems with an empirical study validating clients' banking events behind an unpaid loan. Business organizations may use Gomphy to confirm whether some potential problems hidden in Big Data are against a business goal or not. Gomphy would help find real business problems and improve business value, especially in Big Data and Machine Learning (ML) projects. Four main technical contributions were: 1. a domain-independent Gomphy ontology, helping avoid omissions and commissions in modeling categories of essential concepts and relationships, 2. a method of modeling a concept in a problem hypothesis as a problem hypothesis entity, 3. a data preparation method, supporting to identify relevant features to test in a database and build a dataset; 4. an evaluation method detecting the positive and negative relationships among problem hypotheses and validating the problem hypothesis with feature importance and reasoning scheme.

Future work includes an in-depth study about the impact of correlated and time-series events on validating problem hypotheses, exploring potential solutions to mitigate the validated problem using ML and a goal hypothesis, and developing a reliable Gomphy assistant tool.

References

1. Ross, D.T., Schoman, K.E.: Structured analysis for requirements definition. IEEE Trans. Softw. Eng. **SE-3**(1), 6–15 (1977)
2. Nuseibeh, B., Easterbrook, S.: Requirements engineering: a roadmap. In: Proceedings of the Conference on the Future of Software Engineering, pp. 35–46 (2000)
3. Brownlee, J.: Data Preparation for Machine Learning: Data Cleaning, Feature Selection, and Data Transforms in Python. Machine Learning Mastery (2020)
4. Davenport, T.H., Bean, R.: Big data and AI executive survey (2020). Technical Report, NewVantage Partners (NVP) (2020)
5. Nalchigar, S., Yu, E.: Business-driven data analytics: a conceptual modeling framework. Data Knowl. Eng. **117**, 359–372 (2018)
6. Asay, M.: 85% of Big data projects fail, but your developers can help yours succeed. TechRepublic (2017)
7. Joshi, M.P., Su, N., Austin, R.D., Sundaram, A.K.: Why so many data science projects fail to deliver. MIT Sloan Manag. Rev. **62**(3), 85–89 (2021)

8. Supakkul, S., et al.: Validating goal-oriented hypotheses of business problems using machine learning: an exploratory study of customer churn. In: Nepal, S., Cao, W., Nasridinov, A., Bhuiyan, M.D.Z.A., Guo, X., Zhang, L.-J. (eds.) BIGDATA 2020. LNCS, vol. 12402, pp. 144–158. Springer, Cham (2020). https://doi.org/10.1007/978-3-030-59612-5_11

9. Chung, L., Nixon, B.A., Yu, E., Mylopoulos, J.: Non-functional Requirements in Software Engineering, vol. 5. Springer, Boston (2012). https://doi.org/10.1007/978-1-4615-5269-7

10. Binkhonain, M., Zhao, L.: A review of machine learning algorithms for identification and classification of non-functional requirements. Expert Syst. Appl.: X 1, 100001 (2019)

11. Berka, P., Sochorova, M.: Discovery challenge guide to the financial data set. In: PKDD-99 (1999)

12. Mylopoulos, J., Chung, L., Nixon, B.: Representing and using nonfunctional requirements: a process-oriented approach. IEEE Trans. Softw. Eng. **18**(6), 483–497 (1992)

13. Pearl, J., Verma, T.S.: A theory of inferred causation. In: Studies in Logic and the Foundations of Mathematics, vol. 134, pp. 789–811. Elsevier (1995)

14. Rolland, C., Souveyet, C., Achour, C.B.: Guiding goal modeling using scenarios. IEEE Trans. Softw. Eng. **24**(12), 1055–1071 (1998)

15. Hartmann, S., Link, S.: English sentence structures and EER modeling. In: APCCM, vol. 7, p. 2735 (2007)

16. García, S., Ramírez-Gallego, S., Luengo, J., Benítez, J.M., Herrera, F.: Big data preprocessing: methods and prospects. Big Data Anal. **1**(1), 1–22 (2016)

17. Zheng, A., Casari, A.: Feature Engineering for Machine Learning: Principles and Techniques for Data Scientists. O'Reilly Media, Inc., Sebastopol (2018)

18. Li, J.J., Tong, X.: Statistical hypothesis testing versus machine learning binary classification: distinctions and guidelines. Patterns **1**(7), 100115 (2020)

19. Lundberg, S., et al.: From local explanations to global understanding with explainable AI for trees. Nature Mach. Intell. **2**, 56–67 (2020)

20. Supakkul, S., Hill, T., Chung, L., Tun, T.T., do Prado Leite, J.C.S.: An NFR pattern approach to dealing with NFRs. In: 18th IEEE International Requirements Engineering Conference, pp. 179–188. IEEE (2010)

21. Ishikawa, K.: Introduction to Quality Control. Productivity Press, New York (1990)

22. Vesely, B.: Fault tree analysis (FTA): Concepts and applications. NASA HQ (2002)

23. Supakkul, S., Chung, L.: Extending problem frames to deal with stakeholder problems: an agent-and goal-oriented approach. In: Proceedings of the 2009 ACM symposium on Applied Computing, pp. 389–394 (2009)

24. Molnar, C.: Interpretable Machine Learning. Lulu. com (2020)

25. Ribeiro, M.T., Singh, S., Guestrin, C.: "Why should I trust you?" Explaining the predictions of any classifier. In: Proceedings of the 22nd ACM SIGKDD International Conference on Knowledge Discovery and Data Mining, pp. 1135–1144 (2016)

26. Wille, R.: Restructuring lattice theory: an approach based on hierarchies of concepts. In: Ferré, S., Rudolph, S. (eds.) ICFCA 2009. LNCS (LNAI), vol. 5548, pp. 314–339. Springer, Heidelberg (2009). https://doi.org/10.1007/978-3-642-01815-2_23

27. Abdi, H., Williams, L.J.: Principal component analysis. Wiley Interdiscip. Rev.: Comput. Stat. **2**(4), 433–459 (2010)

28. Reddy, G.T., et al.: Analysis of dimensionality reduction techniques on big data. IEEE Access **8**, 54776–54788 (2020)

Application Track

Application Track

Safety Application Platform of Energy Production Surveillance Based on Data Flow

Xiaohu Fan[1,2] , Mingmin Gong[1(✉)], Xuejiao Pang[1], and Hao Feng[1]

[1] Department of Information Engineering, Wuhan College, Wuhan 430212, China
{9420,9093,9452,8206}@whxy.edu.cn
[2] Wuhan Optic Valley Info & Tech Co., Ltd., Wuhan, China

Abstract. With the development of urbanization, the social demand for energy is increasing. The safety production monitoring of electric power has always been an important issue related to the national economy and people's livelihood. Thanks to deep learning technology, a large number of monitoring and computer vision analysis algorithms have begun to popularize, but only in some simple scenes, or only after investigation and evidence collection. Due to the lack of training samples, the traditional machine learning method cannot be used to train the generative model in hazard situations. Besides, it is a pressure and challenge to the calculation capacity, bandwidth and storage of the system. This paper proposes a platform level solution based on data flow, which can use a large number of cost-effective general-purpose devices to form a cluster, and adjust the task load of each computing unit through software level resource scheduling. The equipment adopts 2U general specification, which can provide better heat dissipation and improve the cooling efficiency of the cluster. At present, the system has been deployed in several pilot projects of the State Grid. It uses LSTM algorithm to establish the contour with normal data training, and uses the deviation of 12.5% as the threshold to identify the abnormal scene. It can accurately identify the obvious suspected abnormal behavior with 98.6% and push it to the operation and maintenance personnel for secondary confirmation.

Keywords: Data flow · Massive video data · LSTM · High performance computing

1 Introduction

With the progress of urbanization, the demand for energy is increasing, and all kinds of equipment are growing in the construction process, involving all production links. Safety production and facility protection are the core work related to the sustainable and healthy development of the industry [1]. The characteristics of energy production and the serious consequences of accidents will directly lead to serious political and economic consequences. Safe and stable operation is of great significance [2].

In recent years, with the popularization of optical fiber technology, video data acquisition of camera and UAV has been widely used in power production safety monitoring system. However, there are not enough people to see the massive data access. Taking

© Springer Nature Switzerland AG 2022
J. Wei and L.-J. Zhang (Eds.): BigData 2021, LNCS 12988, pp. 37–47, 2022.
https://doi.org/10.1007/978-3-030-96282-1_3

China as an example, each province has more than 6000 power transformation and distribution stations on average, and there are more than 40 conventional monitoring cameras at each station, so there will be more than 20000 monitoring probes in each province. Based on the actual monitoring system configuration, the resolution of 1080 pixels and H264 coding of most configured cameras. The amount of data generated by each camera every day is 40 Gb, which can be halved by h265 coding.

Massive data puts great pressure on storage and analysis. For the time being, the video data in most areas can only be used for investigation and evidence collection, so cost-effective solutions are urgently needed by the market [3]. The content of the project is closely related to the safe production of the power industry. Strengthening the monitoring of the production site and its personnel and equipment through in-depth learning of relevant algorithms and ICT technology is the main preventive measure to ensure safe and stable operation.

2 Related Works

2.1 Research Directions

In recent ten years, there have been some serious system intrusions failure in European and American countries, which increased the accidents that serious casualties and have caused great damage to Supervisory Control and Data Acquisition (SCADA) system, so there are many protection works about SCADA system. In China, the power system basically adopts the internal network physical isolation to protect security, and the important networks will not be set with accessible wireless networks.

Due to the development of UAV technology [5], thermal imaging [6] and deep learning technology [7], computer vision can conduct all-round monitoring in many scenes [8]. Scholars have adopted multi-agent model to simulate emergency management. For the situation that the equipment is blocked or visually inconvenient to identify, a large number of sensor and Internet of Things (IoT) monitoring schemes have also been proposed by researchers [9]. The development of GIS and digital twin technology has spawned various emergency simulation systems [10], evacuation systems [11], planning arrangements [12] and material simulation [13], which have strong guiding significance for emergency drills and plans [14] in case of safety production accidents.

2.2 Treatment Methods

At present, all detection methods are mostly based on experience, and there is no accurate instrument for detection in most scenes. Taking the oil leakage detection of oil immersed transformer as an example, the common practice adopted by the power supply company is that during regular inspection, the equipment operation and maintenance personnel use a white flashlight to illuminate the oil leakage points such as the bushing, butterfly valve and gas relay connection of oil immersed transformer and the ground below them, and identify them by "visual inspection" and "nose smell". From the long-term experience of substation inspection, it is found that the efficiency of finding transformer oil leakage by traditional methods is often greatly affected by work experience. When using traditional methods for patrol inspection, it takes a lot of time and has low efficiency, which is not conducive to the completion of daily patrol inspection on time, quantity and quality.

2.3 Hidden Dangerous

In the process of power safety production, the potential safety hazards [15] mainly include the following seven kinds.

Protective equipment is not worn in compliance. For example, basic protective equipment such as safety helmet, insulating gloves, protective clothing, protective eye mask, safety belt and insulating shoes are not worn as required.

Operation error. Due to the large number of equipment in distribution stations and substations, there is the possibility of misoperation.

Staff behavior. On site construction workers sometimes have violations, such as smoking, fighting and dereliction of duty, which may cause hidden dangers to safety.

External damage. For example, vibration may cause safety hazards to nearby equipment.

Equipment failure. Aging and wear of equipment parts, oil leakage of transformer, failure of button or indicator light, etc.

Extreme weather. For example, non-human factors such as rainstorm, sleeping, fire and earthquake have an impact on system safety.

Human factors. Potential safety hazards caused by violent terrorist attacks, hacking, traffic accidents, etc.

Deep learning achieved good accuracy in the field of audio and video analysis. However, deep learning network requires high computing power and bandwidth, and the overall cost of ownership is difficult to control.

3 Solution Architecture

3.1 State Grid Approach

AI platform solutions given by the State Grid would be gradually implemented in the next 3 years. The overall framework is divided into four levels, which is shown in Fig. 1:

Application layer. Call the service layer through the service portal to support the construction of typical application scenarios such as equipment operation and maintenance and customer service.

Service layer. It is composed of algorithms and models, in which model services are also divided into general components and special models.

Platform layer. It includes sample library, model library and AI intelligent platform, which can be divided into training environment and running environment.

Resource layer. It is composed of CPU and GPU computing power, storage and network resources, edge Internet of things agents, and intelligent terminal devices.

3.2 Hardware Deployment

Select qualified substations and other places, supplement necessary equipment such as data acquisition and artificial intelligence model reasoning, and verify the effectiveness

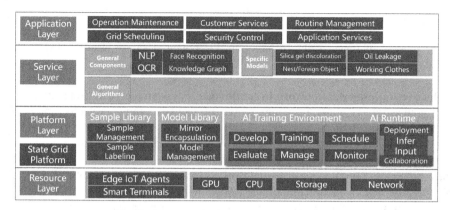

Fig. 1. Framework of AI platform in state grid

of human brain like continuous learning safety production monitoring and detection algorithm in a real environment; Develop demonstration applications around the algorithm model to show the effect of anomaly detection in an intuitive way. The system deployment is shown in Fig. 2. The monitoring data of each position can be collected in real time through the camera, UAV or mobile robot, and then calculated and reasoned at the edge node. Some key data can be returned to the headquarters IDC for training, and the trained model can be distributed to the edge node, which can save bandwidth and computing power and reduce response time. In case of any abnormality, the operation and maintenance personnel shall be notified for handling at the first time.

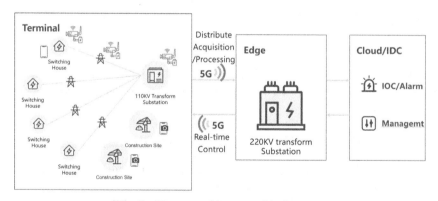

Fig. 2. Abstract architecture of deployment

After calculation, each edge computing node equipped with 2 NVIDIA Tesla T4 GPU graphics card can process 36 channels of camera data. The number of edge nodes is determined by the number of cameras in each substation. Generally, each owns 32 cameras in average, one compute edge node for each substation is well designed. In this way, the monitoring environment for State Grid safety production surveillance in cloud edge collaborative computing is completed.

3.3   Software Architecture

The overall architecture of the system is shown in Fig. 3. The core layer completes the core function process of multi-source data acquisition model efficient reasoning and algorithm continuous learning real-time exception monitoring. The management layer provides different management configurations, which can configure the acquisition, training, reasoning and monitoring modules.

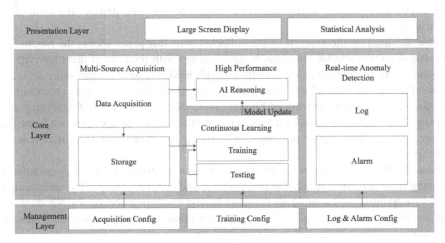

Fig. 3. Software architecture of oil leakage surveillance system

The display layer performs statistical analysis and centralized display of the data generated by the system. The 'acquisition module' in the 'multi-source data acquisition' subsystem is responsible for collecting data from sensors according to the acquisition configuration strategy. On the one hand, it transmits the data to the 'reasoning module' in real time. The 'reasoning module' analyzes the collected data by using the trained model; on the other hand, the data collected by the 'acquisition module' will also enter the 'storage module' for persistent storage, so that the 'training module' in the 'algorithm continuous learning' subsystem can continuously train according to the training configuration strategy, and the latest model after passing the evaluation by the 'evaluation module' will be updated to the 'reasoning module'. The 'reasoning module' will push the results of transformer anomaly detection to the 'real-time anomaly monitoring' subsystem, in which the 'log module' will record all the results for later analysis by the 'statistical analysis module' in the 'presentation layer', and the 'alarm module' will send out alarm events in the analysis results according to the log alarm configuration strategy to remind the staff. Important statistical results and alarms will be visually displayed in the 'large screen display module' to conveniently display the effect of abnormal oil leakage detection of transformer.

Thus a typical cloud edge collaborative environment is completed, the intensive model training work is completed by IDC center node, and the model reasoning calculation is handed over to the edge node.

3.4 Technology Roadmap

According to the above research content, this project adopts the following methods to carry out the research.

Phrase 1. Through the research on the leakage mechanism and historical data of transformer oil leakage, this paper qualitatively and quantitatively analyzes the influence of different factors on transformer leakage from different aspects, as well as various specific appearances caused by it, and on this basis, designs and deploys the basic data acquisition system.

Phrase 2. Based on the analysis of the collected data, this paper adopts the method of anomaly detection similar to human brain learning. With the help of deep learning, it constructs a machine learning model with multiple hidden layers to obtain more accurate and useful features from a large number of training data. The motivation is to establish and simulate the neural network of human brain, which is a method of analyzing and learning by imitating the mechanism of human brain. In the image data, we learn the features that can best express the target, and process the image from the semantic level rather than the pixel level.

Phrase 3. The core structure of anomaly detection algorithm network is completed by using recursive convolutional neural network and convolutional LSTM architecture model for training, and combining the good feature extraction ability of CNN network and the processing ability of RNN, LSTM and other networks for temporal history information.

Phrase 4. According to the environment and historical data of transformer, through repeated training and evaluation, the key parameters such as data update cycle and quantity are determined.

Phrase 5. Explore the use of multi-source data fusion, more comprehensive and more accurate analysis, and further optimize the detection effect.

3.5 Video Content Analysis Technology

In order to analyze the scene behavior of power safety production, it is necessary to accurately capture objects, equipment status, human movement, measure behavior trajectory and identify human posture. The following calculation module based on CV is involved.

Camera calibration. Through the camera automatic correction technology, setting preset position, adjusting angle and focal length, the accurate acquisition of binary image information and the effective conversion between 3D objects are realized.

Environmental correction. When the camera is deployed, the location of the acquisition is fixed, so it is necessary to revise the camera location and restore it to the digital twin of 3D reconstruction.

Equipment status identification. Common scenes include accurate identification of indicator state, meter reading, pressing plate state and switch state.

Personnel Recognition. Judge whether there is a person in the video signal. Firstly, the scattered foreground pixels of the image are aggregated into an object, and the shape and size of the object are analyzed to determine whether it is a human body, and each detected moving human body is marked. This module is often used for access control or suspicious person detection.

Personnel tracking. Once the person is detected and marked, the camera will track the person until they leave the camera coverage. The human motion trajectory output by the module is an important feature root of scene behavior analysis, and is an important part of the whole system design, which needs algorithm linkage camera control signal.

Posture classification. After obtaining the position of the personnel, the head, trunk, limbs, etc. are found on the image according to the model. According to the attributes of the calibration box, the dynamic analysis is carried out on the basis of out of control modeling to identify the posture and whether the key parts are wearing helmets, protective clothing and insulating gloves.

Behavior analysis. According to human motion trajectory and posture recognition, combined with the calibrated area and equipment position after scene correction, human behavior can be judged with the help of specific scene priori. In addition, the key behaviors, such as switching devices or electric shock falls, are also identified.

3D reconstruction. After the processing of the above modules, the front-end IOC display screen presented by the system to the operation and maintenance monitoring personnel is a digital twin system, which can use human-computer interaction to view the required information.

4   Data Flow Scheme

In the aspect of data flow theory, the hardware resources are mapped into data flow abstraction machine, the basic units of resource abstraction are isolated and loosely coupled, and the data transmission between nodes is minimized; Flexible online and offline according to load, providing flow specific dynamic power consumption optimization; Based on fine-grained asynchronous execution mechanism, it provides efficient scheduling of complex business and dynamic load.

In terms of basic resource optimization in multi service environment, hardware resources are reasonably abstracted into fine-grained Processing Element (PE) based on multi granularity and multi types, combined with color pipeline parallel mode plus data parallel mode provided by data flow execution model to form data flow massive parallelism, make efficient use of computing resources and cover up data transmission overhead, Providing maximum parallel efficiency within abstract nodes. Combined with the design goal of data transmission between abstract nodes, it comprehensively provides maximum distributed and parallel computing efficiency.

The advantages of data flow scheme lie in two ways as shown in Fig. 4. On the one hand, the runtime system takes over the responsibilities that cannot be done well by the operating system such as job task scheduling, resource allocation and memory management, and provides the support of parallel mode, processing mode and programming model based on data flow according to the execution model of data flow. On the other hand, The runtime system abstracts the computing units and memory modules of the whole cluster into a unified "abstract machine", provides a unified hardware abstraction for the parallel model, and supports more physical hardware transparently.

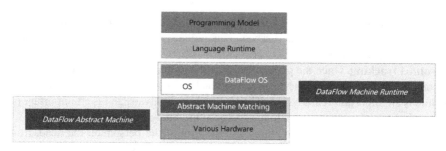

Fig. 4. The abstract layers in Data Flow

5   Implementation and Discussion

In the power safety production monitoring system, the existing business has multiple heterogeneous frameworks from different suppliers. Firstly, massive video data needs to be cleaned and feature extracted. Since there will be more non on demand requirements during service switching, it is necessary to sort out the data and feed it back to different models. After processing, analyze and summarize the results and present them to the original service system. In case of suspected danger, report it in time. The abstract business DAG and work flow is shown in Fig. 5.

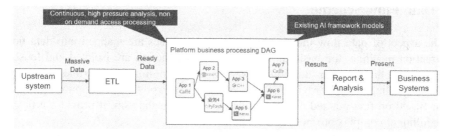

Fig. 5. Business DAG and work flow in safety production surveillance

The business system of artificial intelligence platform based on data flow is divided into four levels, as shown in Fig. 6. The top application layer includes the monitoring, analysis and application of audio and video, the identification of text and indicator signs and status, security applications and externally compatible SDK environment. Then there is the service layer, which contains the sample library of various algorithm operators, fault-tolerant mechanism and dynamic power management. It is the most important application component. The execution layer is the core of data flow, which is mainly responsible for coordinating and scheduling resources to form virtual resources to perform tasks efficiently. The bottom layer is the hardware layer, which is fully compatible with major manufacturers, integrated and flexible, and forms an appropriate IDC according to real-time requirements.

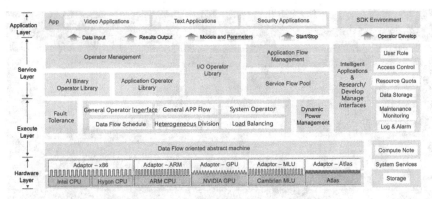

Fig. 6. Screenshot of application system interface

At present, the system has been pilot applied in the State Grid, and the typical scenarios include transmission, distribution, substation and field operation. The real application system is shown in the Fig. 7 and has achieved good results.

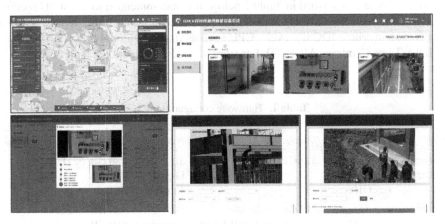

Fig. 7. Screenshot of applications

For typical unsupervised training scenarios, such as external damage, we use the LSTM algorithm to generate a safety profile based on the normal state, and prompt the staff to check if it exceeds the threshold range. As shown in Fig. 8, when a kite flies onto the wire, the AI system has not seen such a scenario, so it is pushed to the operation and maintenance personnel for verification.

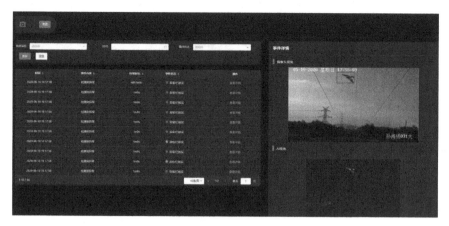

Fig. 8. Anomaly detection application with AI visual angle

For the configuration and selection of hardware, we recommend using conventional general models, which can greatly reduce the customization cost. The current cost-effective scheme is described in Table 1 below. It is recommended to adopt 2U specification, which can achieve better heat dissipation efficiency and further reduce the cost of operation in whole IDC life cycle. CPU such as Intel Xeon or Hygon, GPU such as NVIDIA T4 or Cambrian MLU270. For storage nodes, conventional hard disk array is satisfied.

Table 1. Hardware configurations.

Type	CPU	Memory	AI-GPU	Network	Storage	Specification
Management header note	X86_64 16c 2.5 GHz	512 GB	N/A	10 Gigabit Ethernet	480 GB 960 GB SSD	1U/2U
Compute node	X86_64 16c 2.5 GHz	512 GB	Nvidia T4/MLU270	10 Gigabit Ethernet	480 GB 960 GB SSD	2U/4U
Storage node	X86_64 16c 2.5 GHz	512 GB	N/A	10 Gigabit Ethernet	480 GB SSD + 16*8TB HDD	2U/4U

The data flow scheme combines the high-performance capability of the special model and the flexible architecture of the cloud platform. The specific performance is as follows: 1. Provide task scheduling and resource coordination mechanism facing complex services and dynamic loads to improve the basic computing performance in multi service environment. 2. Reduce the hot spot effect of large-scale business and eliminate bottlenecks, so as to improve the expansion capacity of business and resource scale. 3. Online and near-line QoS scheduling and real-time dynamic power consumption optimization mechanism for online traffic to ensure the stability and sustainability of services. 4. Weaken the demand for dedicated hardware, expand the scope of the supply chain, and explore the optimal solution of cost and power consumption.

References

1. Pliatsios, D., Sarigiannidis, P.G., Lagkas, T., et al.: A survey on SCADA systems: secure protocols, incidents, threats, and tactics. IEEE Commun. Surv. Tutor. 99 (2020)
2. Han, J., Liu, J., Lei, H.: Agent-based simulation of emergency response of urban oil and gas pipeline leakage. In: The 11th International Conference (2019)
3. Karbovskii, V., Voloshin, D., Karsakov, A., Bezgodov, A., Zagarskikh, A.: Multiscale agent-based simulation in large city areas: emergency evacuation use case. Proc. Comput. Sci. **51**, 2367–2376 (2015)
4. Francia, G.A., Thornton, D., Brookshire, T.: Wireless vulnerability of SCADA systems. ACM (2012)
5. Rakha, T., Liberty, A., Gorodetsky, A., Kakillioglu, B., Velipasalar, S.: Heat mapping drones: an autonomous computer-vision based procedure for building envelope inspection using unmanned aerial systems (UAS). Technol.|Archit.+ Des. **2**(1), 30–44 (2018)
6. Danielski, I., Fröling, M.: Diagnosis of buildings' thermal performance-a quantitative method using thermography under non-steady state heat flow. Energy Proc. **83**, 320–329 (2015)
7. Oord, A., Dieleman, S., Zen, H., et al.: WaveNet: a generative model for raw audio. arXiv preprint arXiv:1609.03499 (2016)
8. Yang, L., Shi, M., Gao, S.: The method of the pipeline magnetic flux leakage detection image formation based on the artificial intelligence. In: Proceedings of the International Conference on Video and Image Processing, ICVIP 2017, pp. 20–24. Association for Computing Machinery, New York (2017)
9. Rajalakshmi, A., Shahnasser, H.: Internet of Things using Node-Red and Alexa. In: 2017 17th International Symposium on Communications and Information Technologies (ISCIT), Cairns, QLD, pp. 1–4 (2017)
10. Busogi, M., Shin, D., Ryu, H., Oh, Y.G., Kim, N.: Weighted affordance-based agent modeling and simulation in emergency evacuation. Saf. Sci. **96**, 209–227 (2017)
11. Jumadi, J., Carver, S., Quincey, D.: A conceptual framework of volcanic evacuation simulation of Merapi using agent-based model and GIS. Proc. - Soc. Behav. Sci. **227**, 402–409 (2016)
12. Zhou, C., Wang, H.H., Zhuo, H.: A multi-agent coordinated planning approach for deadline required emergency response tasks. IET Control Theory Appl. **9**, 447–455 (2014)
13. Shafiee, M.E., Berglund, E.Z.: Agent-based modeling and evolutionary computation for disseminating public advisories about hazardous material emergencies. Comput. Environ. Urban Syst. **57**, 12–25 (2016)
14. Cheng, M., Li, Q., Lv, J., et al.: Multi-scale LSTM model for BGP anomaly classification. IEEE Trans. Serv. Comput. 1–17 (2018)
15. Zhu, Q.Z., et al.: Development status and trend of global oil and gas pipelines. Oil Gas Storage Transp. **04**, 375–380 (2017)

Digital Transformation Method for Healthcare Data

Richard Shan[✉] and Tony Shan

Computing Technology Solutions Inc., Charlotte, NC 28277, USA
m@richardshan.com

Abstract. More than 80% of healthcare data is unstructured. The complexity and challenges in healthcare data demands a methodical approach for digital transformation. The Process, Enablement, Tooling, and Synthesis (PETS) method is presented, which provides a holistic approach and discipline to help organizations do it right the first time on digital transformation of unstructured data in the healthcare domain. PETS establishes and evolves a comprehensive knowledgebase for the technology facilitation and implementations in the new era. Details of PETS modules and open-source solutions are discussed. Best practices and real-world PETS applications are articulated in the context.

Keywords: Digital transformation · Healthcare · Unstructured data · Process · Enablement · Tooling · Synthesis · Best practice · Application · Open source · Discipline

1 Introduction

Today's organizations rely more and more on data to make sound business decisions and better their operations. Many healthcare companies go beyond the traditional approach, exploring the opportunities of how to digitize structured data and utilize unstructured data in the domain.

According to International Data Corporation (IDC), in many cases, unstructured text remains the best option for healthcare providers to capture the depth of detail required in a clinical summary, or to preserve productivity by incorporating dictation and transcription into the workflow. Unstructured text records contain valuable narratives about a patient's health and about the reasoning behind healthcare decisions [1].

Unstructured data refers to the information without a pre-defined data model or not organized in a pre-defined format. Examples of unstructured data include books, journals, documents, metadata, health records, audio, video, analog data, images, files, and unstructured text such as the body of an e-mail message, webpage, or word-processor document [2]. Healthcare Unstructured Data (HUD) refers to the information in the formless format in the healthcare domain. There are many sources of HUD, some of which are listed in Fig. 1.

For example, neuroimaging uses various techniques to either directly or indirectly image the structure, function/pharmacology of the nervous system. In Magnetic Resonance Imaging (MRI), magnetic fields and radio waves are used to generate images of brain structures in two or three dimensions, which are unstructured in nature.

© Springer Nature Switzerland AG 2022
J. Wei and L.-J. Zhang (Eds.): BigData 2021, LNCS 12988, pp. 48–63, 2022.
https://doi.org/10.1007/978-3-030-96282-1_4

Fig. 1. Sources of healthcare unstructured data

Enterprises are storing petabytes of unstructured data [3]. A cross-industry consensus shows that 80–90% of all healthcare data remains unstructured [4]. The estimate from the Health Story Project shows that some 1.2 billion clinical documents are produced every year in U.S., 60% of which contain valuable patient-care information "trapped" in an unstructured format [5].

A survey by Healthcare IT News revealed that healthcare organizations are still in a reactive mode as they develop or apply strategies to bring this vast amount of unstructured data in alignment with their Electronic Health Record (EHR) systems. The survey report indicates that 40% of respondents are implementing a solution for unstructured clinical data, 25% are creating a strategy or planning, and 18% are looking at what solutions are available in the marketplace [6].

Gartner points out that mapping and analyzing unstructured data are key imperatives for IT management. But 45% of organizations surveyed have no management tool in place to help better understand the unstructured data [9].

Another study by Quest Diagnostics and Inovalon indicates that 65% of healthcare providers do not have the ability to view and utilize all the patient data they need during an encounter, and only 36% are satisfied with the limited abilities they have to integrate big data from external sources into their daily routines [7].

In the related work previously, Adnan et al. explored the role and challenges in HUD [9]. Anusha presented an approach to transforming HUD to healthcare structured data (HSD) [10]. Yue discussed how to use Hadoop and Drill to archive and retrieve HUD [11]. Gudu reviewed the semantic design for HUD [12]. Similarly, Cambria et al. attempted to apply semantics and sentics with the cognitive and affective information to bridge the gap between HUD and HSD [13]. Further, Giannaris et al. employed an RDF-based method to uncover implicit health communication episodes from HUD [14]. Letinier et al. made use of a knowledge database and gradient boosting trees for HUD [15], and Kalaivanan et al. exploited the natural language processing (NLP) approach to mine HUD [16].

The paradigm has shifted. We are now facing the unprecedented challenges and enormous complexity of today's dynamic world. Traditional methods, despite succeeding to some extent in the past, are no longer effective or good enough. It is not uncommon for some firms and service providers to stick with the old "garbage in, garbage out" approach in attempt to solve new problems in vain. Nowadays, the reality is becoming "transform or die".

2   Methodical Approach

In our view, it is imperative to take a different approach for HUD in the new era, due to the immaturity, complication, inconsistency, fragmentation, and compliance in this domain. It is hard to jump start your HUD journey quickly on the right foot. It is even harder to progress effectively and stay focused on the right track along the way without sound methodology.

We must take cultural changes to embrace the new mindset for transformation and innovation in the flat world. Businesses that miss this crucial window of research and investment run the real risk of being left behind or displaced by more forward-thinking organizations [17].

From a technical perspective, we advocate a pragmatic method to implement HUD solutions, composed of 4 constituents: **Process**, **Enablement**, **Tooling**, and **Synthesis** (PETS) Module, as shown in Fig. 2.

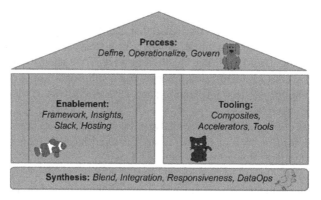

Fig. 2. PETS method

The **Process Module** defines a series of steps and decisions involved in how HUD processing and analysis are conducted as a project execution roadmap.

The **Enablement Module** describes key technical capabilities, addressing core design concerns and runtime aspects for implementations and facilitation systematically.

The **Tooling Module** prescribes a list of packages, tools, accelerators, and patterns developed for higher quality, lower cost, and better time-to-market in HUD application construction, deployment, and operations.

The **Synthesis Module** specifies the in-depth knowledge and expertise on leading-edge innovations and competency in applying various technologies to the healthcare domain solutions in practice.

PETS aims to tackle the HUD roadblocks holistically by formulating a step-by-step routine with the enabling elements to strategize the process. PETS also operationalizes the implementation with right toolsets and empowers practitioners via best-practice disciplines. PETS paves the way to fast track design and development of HUD applications.

PETS promotes standards, open source, objectivity, cloud-native, modularity, compatibility, and repeatability. We will drill down to these 4 modules in greater detail in the subsequent sections.

3   Process Module

The Process Module unifies the conventional heavyweight and lean methods to prescribe hybrid undertakings, providing progressive tasks to implement a HUD analysis solution end-to-end. The module is comprised of 3 major stages: **Define**, **Operationalize**, and **Govern** (DOG). These DOG stages are logical phases in the HUD blueprint, as depicted in Fig. 3.

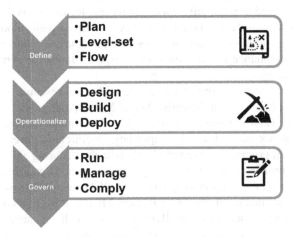

Fig. 3. DOG process module

3.1   Define

To get started, it is pivotal to establish strategic objectives and build consensus on right expectations among the stakeholders. We need to identify the most important and useful sources of data for the organization and focus on what is most relevant. We must determine the outcome of unstructured data analysis, and what is expected to be presented. We can then create detailed specification of the requirements, both functional and non-functional, and categorize the demands in short-, mid-, and long-term groups. We have to set the right priorities on individual items and resolve conflicts with risk mitigation strategies. We need to find the low hanging fruits to tackle first for quick wins in the initial iteration.

3.2   Operationalize

Data is the new oil. It is vital to treat HUD as assets. We ought to take a deeper look at the primary and secondary data in place and on the horizon for the orchestration pipeline, and then construct a solid technical foundation with the functional components for the end-to-end data processing. We shall bake off various competing options to establish a platform, and further, design the parallel data ingestion, for both batch and streaming in real time. We are compelled to frame an extensible data lake for different types of incoming data, followed by a good mix of NoSQL and other stores for refined data and processed information. ELT with robotic processing automation should be considered and developed carefully.

3.3 Govern

Various data analysis is performed with the techniques in statistical modeling, data mining, and machine learning (ML). The output is represented in visual formats, and the results are delivered through different channels and devices as needed. The runtime environment is continuously monitored, and SLA is enforced with automated tools and proactive alerting, resulting in quality services with scalable and reliable clustering and cloud systems. Compliance is enforced with enterprise standards, corporate policies, and regional regulations.

New and revised roles are introduced to the traditional resource model for HUD, such as Chief Data Officer, Data Scientist, Social Media Analyst, etc. Various Centers of Excellence (CoE) and solution workgroups should be formed to develop and nurture best practices in new capabilities and competency. Overarching governance and technology management programs are required to handle the maturing and evolving nature of overwhelming proliferation of cutting-edge technologies. Organization realignment is needed to cope with the cross-cutting aspects of HUD enterprise-wise.

From a lifecycle standpoint, the DOG Process Module leads to a consistent format for planning, organizing, and running HUD projects. It helps manage various project parameters, such as roles and responsibilities, financial budget, resource skillsets, teaming, duration, timeline, requirement tracing, standards, tools, quality control, productivity, security, compliance, dependencies, and risks. Each stage is further broken down to sequential steps in great depth, as illustrated in Fig. 4.

The activities in the workflow are evolutionary in an incremental fashion. Selected steps are chained in time-boxed iterations, e.g., rapid prototyping in phases. Agile techniques are applied in the mini cycles of related steps, like spikes and sprints. Complex steps are further zoomed in to form procedures.

As an example of DOG implementation, MLFlow [18] provides an open-source ML lifecycle platform that manages the exploration, repeatability, installation, and registration. The Tracking function has UI and API for metrics, code versions, parameter comparisons, and others. The Projects component stores the ML code in a reusable and verifiable format that can be shared with other data scientists or used in production. The Models section manages models from various ML libraries, deployed to different serving and inference platforms. The Model Registry handles the model versioning, stage transitions, and annotations collectively.

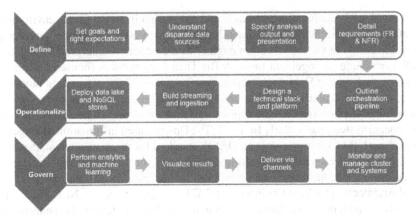

Fig. 4. Steps in DOG process module

4 Enablement Module

The Enablement Module consists of the following pillars: **Framework**, **Insights**, **Stack**, and **Hosting** (FISH), as represented in Fig. 5.

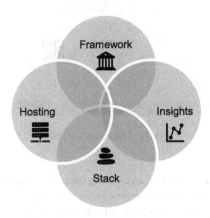

Fig. 5. FISH enablement module

4.1 Framework

It is critical to make full use of comprehensive frameworks to manage various components and processes in a HUD solution. A sophisticated architecture framework consists of multiple viewpoints and views to describe various design and development considerations and concerns. It helps establish a common and consistent practice.

In general, there are four types of frameworks: Big Data, Unstructured Information, Generic, and Specific, as portrayed in Fig. 6. A Generic framework deals with the general hurdles of data handling, storage, and integration in an enterprise environment, like Zeta Architecture, which includes a distributed file system, real-time data storage, a pluggable compute model/execution engine, data containers, enterprise applications, as well as resource management tools. A Big Data framework is focused on Big Data processing, such as Big Data Architecture Framework (BDAF), comprising models for domain, enablement, platform, and technology. SMART Health IT is an open, standard-based technology platform to create applications that seamlessly and securely run across the healthcare system, particularly for EHR. The Unstructured Information Management Architecture (UIMA) framework is an OASIS standard for the development, discovery, composition, and deployment of multimodal analytics for the analysis of unstructured information. A Specific framework is tailored to particular needs and purposes. The General Architecture for Text Engineering (GATE) is a framework for NLP. The Clinical Text Analysis and Knowledge Extraction System (cTAKES) is designed to analyze clinic notes.

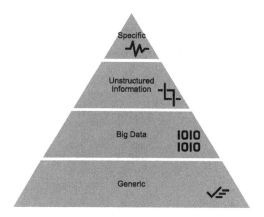

Fig. 6. Framework types

It is preferred to take advantage of an existing framework for a baseline if possible and combine/expand it as needed to be customized for individual needs in HUD processing. If no direct fit can be found, you may have to build one yourself; but it is time-consuming and resource-intensive to construct your own framework from square one. Without prior experience or right resources, this kind of effort could become a high-risk daunting task that makes your team lose precious time and opportunity.

4.2 Insights

Today's era mandates the gathering of all kinds of data available from different channels for business operations. More importantly, we need to examine large data sets collected, containing a variety of data types, to uncover the hidden patterns, unknown correlations, market movements, industry trends, customer behaviors and other useful business information. It is our duty to employ a combination of descriptive, diagnostic, predictive, and prescriptive analytics to extract greater insights, via open-source analytics engines like R and RapidMiner. Moreover, artificial neural networks should be put to good use for deep learning in sentiment analysis, image search, ERP, risk detection, emotion recognition, and so on, to produce more actionable results.

For example, an advanced Big Data analytics platform like KNIME is for open-source cloud-based data exploration and guided analysis, delivering real-time analytical information and predictive/prescriptive models for better decision making. It moves away from traditional business intelligence reporting to more actionable insights, by building a machine learning mechanism that understands the behavioral aspects of decision making. Data events are used to produce semantic information based on rules/patterns for identifying events of significance. The exploratory data discovery via Generative AI produces new insights on new medicines and treatment.

4.3 Stack

Open-form technology stacks bring about freedom, flexibility and extensibility for development and rollout of HUD solutions without lock-ins. Leveraging a half-baked structure of integrated features can speed up solution development in a more agile and open fashion, rather than building from scratch. For instance, Hadoop gives a matured ecosystem for a distributed data lake and corresponding data processing, in conjunction with other analysis products for messaging, streaming, discovery, refinery, analytics, and visualization (e.g., Kafka, Storm, Flink, Spark, Elasticsearch, and Lumify). It is essential to select the best route for an end-to-end platform in open source to make your HUD transformation smoother and faster – adopt, customize, extend, merge or construct one from the ground up.

A well-designed AI platform copes with a large set of cognitive computing services for the development of digital virtual agents, predictive systems, cognitive process automation, visual computing applications, knowledge virtualization, robotics, and drones. The platform is developed using machine learning, natural language processing, genetic and deep learning algorithms, semantic ontologies, pattern recognition and knowledge modelling technologies to provide solutions that deliver cognitive enhancement to experience and productivity, accelerate process through automation and at the highest stage of maturity reach autonomous abilities. For instance, the Machine Learning Canonical Stack (MLCS) by AI Infrastructure Alliance (AIIA) is intended for an abstract AI/ML factory in a plug and play mode, with agnostic orchestration systems, clean API layers, and well-defined communications standards.

4.4 Hosting

The runtime environment for HUD requires scalability and performance to process large-scale volume and diverse types of unstructured data in a timely fashion. Companies are forced to formulate a good combination of private, public and hybrid clouds with Software, Platform, and Infrastructure as a Service (SaaS/PaaS/IaaS).

On-premises hosting moves towards standard-based infrastructure like OpenStack. Database as a Service (DBaaS) supplies faster and more scalable instances of DBMS. Likewise, Hadoop as a Service (HaaS) spins off fast-provisioned clusters in a matter of minutes. For off-premises hosting, subscription-based services are available from various vendors like AWS, Azure, GCP, and Bluemix. Hybrid cloud furnishes a good mix for data spikes, backup/archiving, data exploration, experimentation, and dev sandbox. In addition to Infrastructure as a Service (IaaS) as a hosting solution, companies are obligated to consider PaaS and SaaS models more seriously for their own products/services as well as for supporting functions in the enterprise. Organizations must embrace multicloud models in their cloud strategy and migration roadmap.

In a nutshell, the FISH Enablement Module consists of key disciplines for effective implementations of HUD solutions with higher quality. It seeks to fix commonly occurring issues within a given context, and presents guidance of applying repeatable techniques, to avoid reinventing the wheel. The artifacts comprise reusable patterns to help build HUD data fabric faster and easier. They also promote the consistency in development projects, elevating the communications, productivity, and supportability.

5   Tooling Module

A rich suite of assets is developed and evolved, which are used and reused in numerous projects to accelerate the design, construction, migration, and release of HUD solutions for a number of companies. The Tooling Module has a broad set of **Composites**, **Accelerators**, and **Tools** (CAT).

5.1   Composites

Composites are the ensembles of pre-built and pre-configured units and packages to stitch together reusable pieces via a wizard in a visual format to compose a business or technical solution in a no/low-code mode. They are quickly instantiated, provisioned and deployed to production clouds in composable architecture, in seconds or minutes rather than weeks/months in the traditional way.

For example, a document analytics platform digitizes document-driven operations with a view to automate and simplify the processing. This platform has been designed around core modules such as OCR, entity extraction, NLP, ML, pattern recognition, and robotic processing automation. It converts scanned images/PDFs into digitized formats, which can be easily managed, processed, accessed, and searched. The platform recognizes and extracts key information from both structured and unstructured formats. The solution automates the manual processes of healthcare organizations such as information aggregation, extraction, and verification. It contains key processes for KYC verification, document classification, claim statements, quote digitization, data entry, patient records, admission paperwork, prescriptions, and processing for treatment approvals. OpenHIE is another standard that normalizes data exchange.

5.2   Accelerators

Accelerators are self-contained connectors or building blocks packaged to address the specific data processing needs at relatively lower level, e.g., data types, operations, integration, conversion, etc. They are reusable in a plug-and-play fashion. They help design, develop, integrate, and implement HUD solutions rapidly.

For instance, the openEHR Clinical Knowledge Manager (CKM) is a free, online clinical knowledge resource that promotes an open and multinational approach to clinical informatics, under a Creative Commons license. CKM manages the clinical knowledge in a library of clinical knowledge artifacts, which are mostly openEHR archetypes and templates at the moment. Users interested in modeling clinical content can utilize CKM to contribute to the design and/or enhancement of an international set of archetypes, as the foundation for interoperable EHR. Similarly, Multi-Level Intermediate Representation (MLIR) makes it easier to construct AI applications by addressing the difficulty posed by increased software and hardware fragmentation. It introduces new infrastructure and a design philosophy that make it possible to express and execute machine learning models reliably on any type of hardware.

5.3 Tools

Tools are items used by professionals to create, test, deploy and maintain data services, applications, and systems in Systems Development Life Cycle (SDLC), which are not consumed in the process. Tools may be discrete programs executed separately, or parts of a single large program like IDE. Tools increase the team productivity, sharing, documentation and communication efficiency.

As an example, an Electronic Data Interchange (EDI) document tracker is developed to integrate EDI Applications and help business users to trace the EDI transactions through the EDI dashboard. It also has advanced search function via Elasticsearch to assist flexible lookup of transactional EDI data through correlations based on unique identifiers like PO#, ASN#, ISA ID etc. and retrieve data and their status. The tracking tool improves the visibility of message loss and connectivity and empowers the users. In contrast, OpenBytes makes open datasets more available and accessible, and create an open data standard and format for the AI community.

6 Synthesis Module

The Synthesis Module is focused on the best practices and execution of healthcare initiatives: **Blend, Integration, Responsiveness**, and **DataOps** (BIRD), as described in Fig. 7.

6.1 Blend

Various emerging technologies tend to converge in a ubiquitous manner. We ought to synergize and unify different technologies to create a best-in-class combo. Systems are required to be constructed as suites of independently deployable services in the componentization style via microservices patterns like API Gateway. NLP, ontology, text analytics, knowledge discovery and machine learning are critical ingredients for HUD analysis to harvest unstructured data to reveal semantic relevance in the context. Health Bots make the most of AI and NLP to enable a more natural interaction between humans

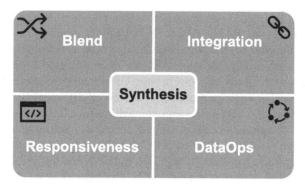

Fig. 7. BIRD synthesis module

and systems, for appointment scheduling, billing/payment processing, and clinical record management.

Connected Health empowers doctors for remote monitoring, diagnosis, and treatment. The blockchain technology is employed for better HUD management, like patient records, treatments, and insurance transactions. 3D-printers are promising in order to revolutionize medical prescription filling and refilling. VR/AR/MxR continues to bridge the cyber and physical worlds, presenting physicians with compound and simulated inside-out views in surgical procedures. Internet of Things (IoT) will generate more unstructured data and connect more medical devices and equipment for better health care and management of metahealth in the metaverse ecosystem.

6.2 Integration

We are obliged to increase loose coupling in complex enterprise integration. Cloud-based integration products and iPaaS are matured to link systems in enterprise data centers and public clouds. In addition to the enterprise solutions like MuleSoft and Apigee, connectors are available to connect consumer products, like IFTTT and Zapier. Containerization boosts the interoperability for seamless integration, such as Docker, Mesos, and Kubernetes. In the IoT environment, field and edge gateways (e.g., Eclipse Kura) facilitate the edge and fog computing for the medical equipment and wearables like Fitbit fitness trackers.

A data integration platform is an integrated solution for capturing & managing data and generating actionable insights through advanced analytics to offer price, performance, and time benefits. It uses agile methodology to operationalize the insights in the quickest possible time, built with open source and therefore minimized license cost. It uses an OpEx model, with no CapEx involved, in the container deployment. It comes with ready-to-use infrastructure on major public clouds, completely scalable up or down with provision to add or reduce nodes.

For example, the Open Integration Hub offers an open-source framework for easy data synchronization between business applications. It comprises standard data models, connectors, platform, integration flows, and guidelines. Jitsu supplements a fully scriptable data ingestion engine, collecting data from 140+ sources to build accurate profile of users, and automatically resolves user geo-location based on IP address. On the other hand, Airbyte gives an open-source data integration engine that consolidates data in warehouses, lakes, and databases, and commoditizes data integration. It comes with more than 100 pre-built ELT connectors, Connector Development Kit, and reverse-ETL connectors.

6.3 Responsiveness

It is needless to say that your user interactions must be enriched and optimized to enhance the customer experience and functionality. Responsive design is essential to capitalize on the growing mobile market. Modern systems are built to scale the contents and engage users cohesively on all channels and devices. Personalized user experience extends customized campaigns and a relevant engagement across digital channels, resulting in improved loyalty. Reactive programming can be leveraged to deal with the static and dynamic data flows for the propagation of change. Matured tools like Meteor should be used to handle the tough issues encountered in building rich front-end solutions with Single Page Application (SPA), e.g., state management and memory leaks in a large system. Portable cross-device support for native platforms becomes a necessity for omnichannel interactions.

For example, a digital customer experience management platform offers an open-source data mesh for Experience-as-a-Service, encompassing various modules (linguistics – NLP and text analytics; automation - OCR, Information Extraction, and Machine Learning capabilities; and Analytics - Big Data, social analytics, lead generation, and search relevance) to manage connected customer experiences across channels, devices, and domains. It integrates, transforms and processes cross-channel data to provide personalized and engaged experiences, as highlighted in Fig. 8.

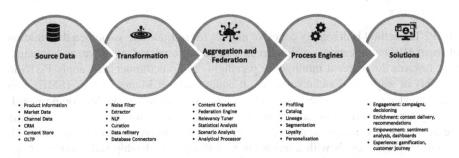

Fig. 8. Digital customer experience management platform

6.4 DataOps

DataOps tackles the key impediments in handling diverse workloads of HUD for quality assurance, Big Data engineering, operations, governance, and collaborations. Continuous everything (CX) becomes a new norm in provisioning and scaling HUD solutions, in continuous integration, testing, deployment, and delivery. Automated QA and configuration improve the productivity and efficiency. Standardized toolchains play a crucial role for quick turnaround and fast experimentation. In particular, Infrastructure as Code (IaC)-based solutions like Terraform can dramatically shorten the deployment time and remove the hefty CapEx to get started.

A DataOps framework is a fully integrated and loosely coupled solution for enterprise data provisioning and data lake management, which is an easy-to-use solution that facilitates data ingestion, enrichment, guided transformation, enhanced monitoring, faster deployment with automated workflow generation and metadata management. It simplifies and automates common data management activities to help enterprises focus time and resources on leveraging business-driven insights and analytics. This improves time to market, implementation efforts, and competitive advantages.

As an example, Meltano presents a DataOps platform in open source, which integrates best-of-breed extensible technologies and tools in the data lifecycle: Model, Extract, Load, Transform, Analyze, Notebook, Orchestrate. The platform entails a library of over 200 connectors with the Singer taps and targets standard for extractors and loaders, leveraging dbt for transformation, Airflow for orchestration, and Superset for visualization. Meltano streamlines build, configuration, deployment, and monitoring, allowing engineering teams to take use of best practices such as version control, code review, and continuous integration and delivery. Databand complements this with a data observability platform that allows data engineers to see if data is available, consistent, and clean as it moves from source to pipeline in a proactive approach. Its open-source library enables users to track data quality, monitor pipeline health, and automate advanced processes via custom metrics and data health indicators.

7 PETS Usage

7.1 Practice

The PETS method is driven by cohesive models to facilitate and smoothen the buildout of smart HUD applications, which become more personalized, adaptive, and anticipatory, towards the ambient intelligence paradigm. The practitioners can apply PETS in a variety of ways for domain solutions. It equips Subject Matter Experts (SME) with deep knowledge and real-world project guidance for designing and developing world-class systems in various organizations. It is purposely built for operating and running the most complex enterprise and cloud environments, on-premises or off-premises. It empowers professionals with a unique mix of thought leadership and hands-on practices in large-scale transformations with innovation and incubation of disruptive HUD solutions.

In practice, PETS aligns with the work of Semantic Networks, part of which is instantiated in Unified Medical Language System (UMLS), developed and maintained by the United States National Library of Medicine. In the biomedical sciences, UMLS is a compendium of numerous controlled vocabularies. It enables translation between various terminology systems, with a mapping framework between different vocabularies. It is also regarded as a comprehensive thesaurus and ontology for concepts in biomedicine. UMLS includes NLP capabilities, aimed mostly at the informatics systems professionals.

Whether it is Current Procedural Terminology (CPT), International Classification of Diseases (ICD), Diagnosis-Related Group (DRG), or their relationships, UMLS has it all, with quarterly updates, ready to be used for free. It is also accessible to external systems via open APIs, allowing intelligent systems to be built using the information accumulated in the system.

The analytics core of the future healthcare information ecosystem provides close collaborations among all information silos and functional systems, with knowledge hubs similar to UMLS as a key enabler. The analytics core consists of clinical decision support systems, NLP, and correlated rules and heuristics that analyze patient, medication, and compliance knowledge in order to provide suitable responses to inquiries from diverse stakeholders including doctors, administrators, and others involved in the care delivery process.

We must use effective analytics to make sound decisions and take appropriate actions based on a holistic view of integrated, relevant, and timely data. We will be able to make a huge positive impact on our capacity to provide better healthcare results with reduced costs and increased patient happiness at the highest levels of compliance, only if we have the right information, at the right time, for the right person.

7.2 Applications

We leveraged PETS to drive the development of a customer complaint analysis tool for mining unstructured text in health-app chats, as a case study. The online conversations between the users and customers agents are captured and stored. Regular chat records are fed to the data lake daily. The input data is parsed via NLP and stored in a Hadoop cluster. Aggregation and refinery are conducted to store the summarized info in a NoSQL data repository. Sentiment analysis and statistical models are performed in R, with the reporting and visualization presented by the Pentaho tools. Data streaming is enabled to analyze critical data events and signals for near real-time action recommendations. Selected chats flow to the analysis platform continuously in stream, based on the business rules and triggers. Web-based interfaces allow admins, users, and modelers to manage and interact with the systems and clusters. PETS shortened the development cycles and reduced the overall project cost significantly.

Another solution example is an application for the social media insights at a leading healthcare giant. We use the PETS accelerators and connectors to first crawl data from web, based on Key Opinion Leaders (KOL) list. Then we categorize KOLs from the online interactions: activity, relevance, and engagement. We analyze the frequency of posts, segmentation and clustering of comments, relevance of opinions, reaction of community, engagement with affiliates, and trends. Subsequently, we customize the contents and channels, depending on the topics and interests for KOLs. PETS helps speed up the time to market with reusable assets and reference architecture.

Due to the space constraints, details of the implementations are provided in separate publications.

8   Conclusion

Many healthcare organizations are adopting and leveraging emerging technologies to revamp their data platform architecture and embed analytics in processes and applications for unstructured data. Advanced technologies enable healthcare systems to find actionable insights from all the data available at their disposal, to make improved decisions and gain competitive advantages.

It is not only important but also necessary to do it right the first time (DIRFT) in the new paradigm with tremendous challenges and unparalleled complexity. We must leverage proven methodology and best practices to avoid a false start and strategize a winning path to succeed. A methodical approach is a must to guide us through the digital transformation journey systematically: Process, Enablement, Tooling, and Synthesis (PETS). Each module consists of various components to address the design and development concerns, pain points, and obstacles in HUD implementations.

The PETS method offers a cross-disciplinary framework with a rich set of workflows, platforms, patterns, tools, accelerators, and practice guidelines to dramatically speed up the HUD implementations for small, medium, and large size organizations. The method can effectively deal with the issues and headaches in HUD. It is also applicable to handle the barriers and constraints in the general Big Data space and beyond, such as IoT, Edge Computing, Blockchain, etc.

References

1. Feldman, S., Hanover, J., Burghard, C., Schubmehl, D.: Unlocking the Power of Unstructured Data. IDC Health Insights (2012)
2. Biswas, R.: Introducing data structures for big data. In: Singh, M., Kumar, G.D. (eds.) Effective Big Data Management and Opportunities for Implementation, p. 50. IGI Global (2016)
3. Komprise: The 2021 Komprise Unstructured Data Management Report (2021). https://www.komprise.com/wp-content/uploads/Komprise-State-of-Unstructured-Data-Management-Report.pdf. Accessed 1 Nov 2021
4. HIT Consultant: Why Unstructured Data Holds the Key to Intelligent Healthcare Systems (2015). http://hitconsultant.net/2015/03/31/tapping-unstructured-data-healthcares-big gest-hurdle-realized. Accessed 1 Nov 2021
5. GE Healthcare: Taming the Data Monster – the right step towards meaningful analytics (2016). http://www.ge-health-it-views.com/ge-health-it-views/taming-the-data-monster-the-right-step-towards-meaningful-analytics. Accessed 1 Nov 2016
6. Fujitsu: Achieving Meaningful Use Interoperability. Healthcare IT News (2011)
7. Gartner: Does File Analysis Have a Role in Your Data Management Strategy? (2014). https://www.gartner.com/en/documents/2677319/does-file-analysis-have-a-role-in-your-data-management-s. Accessed 1 Nov 2021
8. Quest and Inovalon: Finding a Faster Path to Value-Based Care (2016). https://www.ahip.org/wp-content/uploads/2016/07/Study-Finding-a-Faster-Path-to-Value-Based-Care.pdf. Accessed 1 Nov 2021
9. Adnan, K., Akbar, R., Khor, S.W., Ali, A.B.A.: Role and challenges of unstructured big data in healthcare. In: Sharma, N., Chakrabarti, A., Balas, V.E. (eds.) Data Management, Analytics and Innovation. AISC, vol. 1042, pp. 301–323. Springer, Singapore (2020). https://doi.org/10.1007/978-981-32-9949-8_22

10. Anusha, R.: Novel approach to transform unstructured healthcare data to structured data. Int. J. Res. Appl. Sci. Eng. Technol. **9**, 2798–2802 (2021). https://doi.org/10.22214/ijraset.2021. 36972

11. Yue, H.: Unstructured healthcare data archiving and retrieval using hadoop and drill. Int. J. Big Data Anal. Healthc. **3**, 17 (2018). https://doi.org/10.4018/IJBDAH.2018070103

12. Gudu, J., Balikuddembe, J., Mwebaze, E.: Semantic design for unstructured health data: a review, pp. 975–980 (2017). https://doi.org/10.1109/CSCI.2017.168

13. Cambria, E., Hussain, A., Eckl, C.: Bridging the Gap Between Structured and Unstructured HealthCare Data through Semantics and Sentics (2011)

14. Giannaris, P., et al.: RDF-Based Method to Uncover Implicit Health Communication Episodes from Unstructured Healthcare Data (2020)

15. Letinier, L., et al.: Artificial intelligence for unstructured healthcare data: application to coding of patient reporting of adverse drug reactions. Clin. Pharmacol. Ther. **110** (2021). https://doi. org/10.1002/cpt.2266

16. Kalaivanan, S., Reshmy, A.K.: Unstructured big data in health care using natural language processing. Res. J. Pharm. Biol. Chem. Sci. **7**, 230–237 (2016)

17. Miklovic, D.: Digital Transformation Means Using Data to Enable New Approaches (2016). https://dzone.com/articles/digital-transformation-means-using-data-to-enable. Accessed 1 Nov 2021

18. MLFlow. https://mlflow.org. Accessed 1 Nov 2021

Citizens' Continuous-Use Intention to Open Government Data: Empirical Evidence from China

Hui Jiang[1,2(✉)], Yaoqing Duan[2], and Yongdi Zhu[2]

[1] School of Management Science and Engineering, Shandong Institute of Business and Technology, YanTai, China
201513103@sdtbu.edu.cn
[2] School of Information Management, Central China Normal University, Wuhan, China

Abstract. An improved understanding of the factors that influence citizens' continuance use intention will help to promote and improve the practice of open government data. This paper constructs an integrated model that provides insight into factors that influence citizens' continued intention to use open government data. The model contains 296 effective samples from questionnaires, which are then tested by the Structural Equation Model. It is found that perceived usefulness and satisfaction significantly affect the public's continuous adoption of OGD; expectation confirmation significantly affects satisfaction and perceived usefulness, and thus indirectly affects the public's continuous adoption of OGD; perceived ease of use significantly affects the satisfaction, trust in government and trust in the Internet significantly affects the expectation confirmation. But perceived usefulness, trust in government and trust in the Internet had no significant effect on public satisfaction.

Keywords: Continuous-use intention · Open data · Open government data

1 Introduction

Open government data (OGD) can produce social benefits and financial value, although this depends on continuous use. In recent years, governments around the world have tried to make open data more accessible and easier to use for the public [1]. Open government data refers to data produced or entrusted by the government or government-controlled entities, which can be freely used, reused and redistributed without restrictions [2]. Those who work in academia and politics generally believe that OGD can encourage social innovation, enhance government transparency [3], achieve economic benefits [4] and develop innovative data applications [5]. In addition, value-added services and related data analysis industries produced by data utilization can reduce the cost of governance and improve government services [6].

This paper is supported by the soft science Project of Shandong Province, China (Grant No. 2021RKY01013).

J. Wei and L.-J. Zhang (Eds.): BigData 2021, LNCS 12988, pp. 64–78, 2022.
https://doi.org/10.1007/978-3-030-96282-1_5

However, there is a lack of knowledge about how often open government data initiatives achieve their objectives [7]. If expected value and benefits through open data are to be realized, then it is essential to achieve the appropriate provision and usage of data [8]. Local government departments currently generally fail to fully consider user's demand when releasing data and do not sufficiently try to encourage citizens to take advantage of open data [9]. Both have resulted in a low public awareness and utilization rate, and the expected benefits and potential value of OGD have not been fully realized.

Although previous research has shown that OGD has great potential value, creating and sustaining sufficient value remains difficult because of government actions [5]. OGD derives from numerous fields and departments and has diverse structures and features. Its distinctive life-cycle and systematic features can result in the environment, scale and quality of open data, in addition to other factors, affecting the data utilization rate. The essence of data utilization is the effective deployment and use of data resources. Researchers have become increasingly preoccupied with identifying factors that influence the public's continued intention to use open data and have tried to promote more effective measures that improve data utilization. Public participation and cooperation have the potential to support government service innovation, and because government departments releasing one-way open data is not the fundamental purpose of open data, continuous use of OGD actually appears to be key to realizing its value. Despite this, contemporary research appears to focus more on providing, rather than using OGD [8].

A few studies have sought to explore the factors that affect the willingness of the public to engage and use OGD [1, 8-11]. However, most of these studies are case studies or qualitative studies, and there is a lack of empirical studies. On this basis, this paper engages the following research questions:

RQ1. Which factors influence and limit public willingness to identify and use OGD?
RQ2. How can public use of OGD be continuously stimulated?

In answering these questions, this paper will build an integrated model that integrates the Expectation Confirmation Model and Trust Model to conduct empirical research. It will identify the factors that influence citizens' continuous use of OGD, and will specifically explore the impact mechanism of the public's continuous use. These determinants can help policy-makers and designers of OGD to identify problems in the process of releasing open data and realize the potential value of OGD. This paper is structured as follows. In the next section, the research model and major hypotheses will be developed, and relevant literature will be reviewed. Section 3 covers the methodology and provides data analysis; Sect. 4 discusses findings, and Sect. 5 highlights the paper's contributions, implications and limitations, and also identifies directions for future research.

2 Theoretical Model and Research Hypothesis

Drawing on the Expectation-confirmation Theory and Trust Theory, this study constructs an improved model to study the influencing factors and mechanisms that influence citizens' continuous use of OGD.

2.1 Citizens' Continuous-Use Intention and OGD

OGD can be freely used and reused by all users. Data users can develop APP applications, data-media integrated products and research reports by applying it. This prevents an ecosystem of OGD from forming and promotes development, value presentation and a virtuous cycle of data ecological chain [12]. Public consumers of OGD are also the key influence that can promote the utilization and value-realization of OGD, and this can encourage government departments to further release data, which improves its data governance and credibility. But the government does not view the utilization of open data as important, and even solely focuses on its release.

This paper therefore studies citizens' continuous-use intention in relation to OGD. It defines continuous-use intention in relation to OGD as the willingness of the public to continue to believe in and use OGD after they have obtained and used it through the OGD platform. This will help to promote its value-added utilization and sustainable development.

More recently, scholars have begun to focus on identifying influencing factors with the aim of encouraging the adoption and usage of OGD. Open data usage is defined as activities related to stakeholder's use of open data [8]. The value of OGD increases with the number of reuses. But knowledge about open data use is scarce [11]. Research shows that 48.3% of 156 OGDIs successfully achieved public use [7].

A few empirical studies from various countries have been conducted to examine the determinants of OGD usage. There is still no consensus on the factors that influence the public continuance intention to OGD and research conclusions differ considerably across individual countries.

Chinese academia mainly focuses on analyzing the current utilization, influencing factors and obstacles that relate to OGD, and few Chinese scholars have addressed the user's utilization behavior and intention. Duan conducts cluster analysis and correlation analysis of the data browsing and downloading rate of the OGD platforms located in six cities across China, and concludes that the utilization status of open data in different themes and cities is unbalanced, and that the overall correlation between browsing and downloading rate is weak [13]. Zhu draws on the SOR framework to explore the user's continuance intentions in relation to government open data platform, and finds that the environmental stimulus and flow conditions of government data releasing platform positively impacts the user's continuance intention [14]. Gao draws on perceived trust and network externality variables to study the initial adoption intention that users have in relation to the government data releasing platform, and specifically refers to the strength of the Unified Theory of Acceptance and Use of Technology (UTAUT) [15]. This research has explored the Chinese public's utilization intention of the OGD in some detail. Their Chinese counterparts, in contrast, have focused on theoretical analysis and qualitative research and have, as a result, conducted less empirical research that provides validated research conclusions.

2.2 Research Model and Hypothesis Development

This paper draws on the Expectation Confirmation Model (ECM) and Trust Model (TM), to construct an intended model for citizens' continuous-use intention of OGD.

Expectation-Confirmation Model. The Expectation Confirmation Model (ECM) is adapted from the Theory of Consumer Behavior, which has attracted domestic and international academic attention and has been applied in various fields. This model is integrated with theoretical and empirical findings from previous Information System (IS) usage research to theorize a model of IS Continuance [16]. A Post-Acceptance model of IS continuance combines ECM with the theory of Information System Adoption; it suggests that users' continuance intention is determined by their satisfaction with IS use and the perceived usefulness of continued IS use. User satisfaction is, in turn, influenced by their confirmation of expectation from previous IS use and perceived usefulness [16].

This research will therefore be based on the Expectation Confirmation Model and it assumes that citizens' expectation confirmation, perceived usefulness and satisfaction will all affect their willingness to continue using OGD. On the basis of this model, this paper proposes the following:

Hypothesis 1: the public perception of the usefulness of OGD will positively impact public satisfaction.
Hypothesis 2: the public perception of the usefulness of OGD will positively impact public continuance intention.
Hypothesis 3: public expectation confirmation related to the utilization of OGD will positively impact public perceived usefulness.
Hypothesis 4: public expectation confirmation after the utilization of OGD will positively impact public satisfaction.
Hypothesis 5: public satisfaction will positively impact public continuance intention related to OGD.

In addition, OGD is difficult for the general public to obtain. Users are entitled to freely download and use it, and government departments are supposed to ensure that this is the case as this will help to improve its ease of use. Availability, data file type, data quality, release type, standard and format and other technical factors are the fundamental norms that the public use to judge OGD, and they also provide a basis for the public to assess the perceived ease of use of OGD. The question whether the OGD is clear, convenient to access for the public and easy to understand will affect the public continuance intention. Domestic government departments currently generally lack standardized databases and a clear understanding of how to release data in a way that is easily comprehensible for users and that can be applied to technological reuse.

This paper's theoretical model incorporates perceived ease of use variables to reflect the complexity and particularity of the OGD, and it proposes the following:

Hypothesis 6: the public perception of the ease of use of OGD positively impacts public perceived usefulness.
Hypothesis 7: the perceived ease of use of OGD positively impacts user's satisfaction.

Trust Model. Scholars also generally believe that trust factors have a key influence on the adoption, satisfaction and continuance intention of e-governance. Public trust in the OGD does not only consist of public trust in the government but also relates to technology, as OGD needs to be released through the network platform. Srivastava and TEO [17] propose the public trust grid framework of e-governance, and suggest that if citizens' trust in government and network is rather high, they will appear to be initiative and synergistic. Belanger and Carter [18] integrate the Technology Acceptance Model and the Diffusion of Innovations and Network Trust Model to construct the Influencing Factors Model of Public Utilization Intention (for e-governance). Jiang [19] suggest that trust in government and technology has a significant positive impact on the public utilization intention for e-government. On the basis of the Trust Model and this research literature, this paper proposes the following:

Hypothesis 8: users' trust in the government positively affects their expectation confirmation.
Hypothesis 9: users' trust in the network positively affects their expectation confirmation.
Hypothesis 10: users' trust in the government positively impacts their satisfaction.
Hypothesis 11: users' trust in the network positively impacts their satisfaction.

Theoretical Model of the Citizens' Continuous-use Intention to OGD. These hypotheses help to establish a theoretical model of the factors that influence the public's intention to continue to engage with OGD. It is shown in Fig. 1.

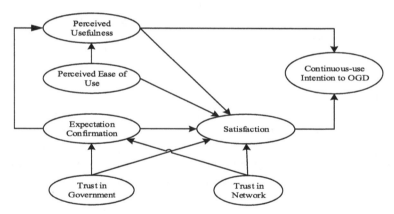

Fig. 1. Theoretical model of the citizens' continuous-use intention for OGD.

This model discusses the direct impact that perceived usefulness and satisfaction have on the public continuance intention, and also highlights the indirect impact of a number of factors (expectation confirmation, perceived ease of use, and trust in the government and network) in this regard. Its application will make it possible to identify the factors that influence the public's constant utilization of OGD. Table 1 defines the model's 7 factors; refers to the variable definitions in the Expectation Confirmation and Trust Model; and combines the features of OGD.

Table 1. Definition and source of model factors

Factors	Definition	Source
Perceived usefulness	Does the user believe that the use of OGD can improve his/her performance	Literature [20]
Perceived ease of use	Does the user believe that OGD is easy to use?	Literature [20]
Expectation confirmation	How does the user's perceived performance of OGD performance compare against his/her initial expectation?	Literature [16]
Trust in government	Does the user believe that government departments provide OGD services with integrity and with a high level of capability?	Literature [18]
Trust in network	Does the user believe that the Internet is a reliable medium that can provide accurate and safe OGD?	Literature [18]
Satisfaction	Does utilization of OGD platform system produce a happy or positive emotional state?	Literature [16]
Continuous-use intention to OGD	Does the user intend to continue to use the OGD after initial use?	Literature [16]

3 Methodology

3.1 Development of Questionnaire

Data obtained through the survey method are used to examine the research hypotheses. The existing mature items, features of OGD and Chinese conditions make it possible to obtain the measurement items of model variables. After the first draft of the questionnaire was completed, 51 graduate students were asked to carry out the pre-test, and 3 professors were asked to provide their suggestions on questionnaire design, which were used to adjust and modify the questionnaire. In order to avoid confusion or boredom, the questionnaire applied the Level 5 Likert scale, including 30 multiple choice questions. The measurement items were modified on the basis of the mature model and combined with the actual situation of OGD in China. Variable measurement items and the literature sources for each of the model's variables are shown in Table 2.

Table 2. Variable measurement items and literature source

Factors	Measurement items	Source
Perceived usefulness	PU1: OGD is very helpful to my work	Literature [20]
	PU2: OGD is very helpful to my life	
	PU3: Using OGD saves me a lot of time	
	PU4: OGD provides me with a lot of valuable information	
Perceived ease of use	EOU1: Learning to use OGD is very easy for me	Literature [20]
	EOU2: OGD is clear and easy to understand	
	EOU3: I can easily find the data I need on the platform of OGD	
	EOU4: I find it easy to access and use OGD	
Expectation confirmation	C1: My experience of using OGD was much better than I expected	Literature [16]
	C2: The data services provided by the OGD platform are much better than I expected	
	C3: Most of my expectations about OGD have been generally satisfied	
Trust in government	TG1: I think I can trust the government	Literature [18]
	TG2: I believe the government has the ability to provide high-quality OGD	
	TG3: I believe the government is concerned about my interests when providing open data	
	TG4: In my opinion, the governments that release data are trustworthy	
Trust in network	TI1: The Internet has enough security measures. I'm very relieved to get data through it	Literature [18] Literature [18]
Trust in network	TI2: I am sure that law and technology can fully protect me from all kinds of network problems	
	TI3: The Internet is now generally a sound and secure environment	

(*continued*)

Table 2. (*continued*)

Factors	Measurement items	Source
Satisfaction	S1: I am very satisfied with the overall experience of using OGD for the first time	Literature [16]
	S2: I am very delighted with the overall experience of using OGD for the first time	
	S3: I was very satisfied with the overall experience of using OGD for the first time	
Continuous-use intention to OGD	CI1: I will continue to use OGD	Literature [16]
	CI2: I plan to stop using OGD and choose other alternative data sources	
	CI3: If possible, I will not continue to use OGD	

3.2 Data Collection

This paper used the Network Survey Method to collect data. When compared with the Traditional Survey Method, it is found to be low cost, highly accurate and also quick. In addition, it also uses fixed samples. Questionnaires were distributed through the online survey tool "Questionnaire Star", which is a professional online questionnaire survey evaluation and voting platform that is focused on collecting data, providing design questionnaires, customizing reports and analyzing survey results. When compared against traditional survey methods and other survey websites or systems, "Questionnaire Star" is fast, easy to use and low-cost, and this is why it is widely used by enterprises and individuals.

A comparison against the Internet user samples from Statistical Reports on Internet Development in China[1] that CNNIC released in 2019 shows that this study's participants were generally older and get benefit from higher education. The proportion of civil servants and public institution workers was also relatively high. Taking into account the fact that OGD in China is still in the initial stage of development, its users should have a certain degree of data literacy and network experience; most were staff members of the government and related institutions, and the number of students under the age of 20 was relatively small.

Further descriptive analysis of sample data was undertaken by applying SPSS20.0, which provided the descriptive statistics in Table 3. These showed that the standard deviation of all variables was less than 1, which confirmed that subjects have relatively consistent opinions, variable dispersion was small and the sample had high data quality.

---

[1] http://www.cac.gov.cn/2019zt/44/index.html.

Table 3. Descriptive statistic

	N	Minimum	Maximum	Mean	Standard deviation
PU1	296	1	5	3.94	.802
PU2	296	1	5	3.94	.789
PU3	296	1	5	3.94	.811
PU4	296	2	5	3.98	.736
EOU1	296	1	5	3.72	.864
EOU2	296	1	5	3.66	.833
EOU3	296	1	5	3.41	.875
EOU4	296	1	5	3.47	.894
C1	296	1	5	3.51	.852
C2	296	1	5	3.52	.863
C3	296	1	5	3.54	.867
TG1	296	1	5	3.94	.840
TG2	296	1	5	4.15	.753
TG3	296	1	5	3.85	.804
TG4	296	1	5	4.01	.807
TI1	296	1	5	3.71	.922
TI2	296	1	5	3.61	.967
TI3	296	1	5	3.43	.936
S1	296	1	5	3.60	.845
S2	296	1	5	3.66	.803
S3	296	1	5	3.53	.831
CI1	296	2	5	4.05	.697
CI2	296	1	5	2.21	.843
CI3	296	1	5	2.05	.877
Valid N (list state)	296				

3.3 Data Analysis

Test on Reliability and Validity. Reliability is the consistency and stability of measurement tools. The ideal reliability measurement tool can truly reflect the attributes of the tested object. Cronbach's Alpha is a common index of this measurement, and its highest value is 1. In practical research, if the coefficient is equal to or more than 0.7, it would generally be considered to be highly internally consistent.

SPSS 20.0 was used to further test the reliability of each latent variable in the questionnaire, and the results are shown in Table 4. The alpha coefficient of each sub-scale was above 0.7, and the alpha coefficient of the total scale values exceeded 0.9, which confirmed the scale was highly reliable.

Table 4. Reliability analysis of the latent variable

Latent variable	Number of measurement items	Cronbach's Alpha	Cronbach's Alpha value based on standardized items
Perceived usefulness	4	0.878	0.878
Perceived ease of use	4	0.86	0.86
Expectation confirmation	3	0.89	0.891
Trust in government	4	0.906	0.906
Trust in network	3	0.881	0.881
Satisfaction	3	0.922	0.923
Continuous-use intention to OGD	3	0.722	0.72

Validity refers to the questionnaire's 'correctness'– that is, the degree to which it can measure the features that will be assessed. Higher validity indicates that the measurement results better reflect the subject's real attributes. Table 5 shows the results of the KMO and Bartlett test after SPSS20.0 was applied to the questionnaire results. The KMO and Bartlett test coefficient was 0.951 and the significance of the Bartlett sphericity test was 0.000, which indicated the questionnaire results are well-suited to factor analysis.

Table 5. KMO and Bartlett test

Kaiser Meyer Olkin measurement of adequate sampling		0.951
Bartlett's sphericity test	Approximate chi square	5294.480
	Df	276
	Sig	0.000

Exploratory Factor Analysis. Table 6 shows the results of exploratory factor analysis addressed to this study's latent variables.

Table 6. Exploratory factor analysis

Variables	Items	Factor loading			
		1	2	3	4
Perceived usefulness	PU1			.790	
	PU2			.787	
	PU3			.817	
	PU4			.761	
Perceived ease of use	EOU1	.626			
	EOU2	.631			
	EOU3	.768			
	EOU4	.747			
Expectation confirmation	C1	.780			
	C2	.726			
	C3	.719			
Trust in government	TG1		.805		
	TG2		.747		
	TG3		.730		
	TG4		.776		
Trust in network	TI1				.784
	TI2				.799
	TI3				.792
Satisfaction	S1	.786			
	S2	.729			
	S3	.718			
Cumulative variance(%)		49.597	57.159	63.735	68.704

Four factors with eigenvalues greater than 1 were extracted, and the variance interpretation rate of the 4 factors was 68.704%, which indicated that potential variables have good interpretative ability.

Confirmatory Factor Analysis. In order to further confirm the reliability and validity of the measurement items, the Maximum Likelihood Method was used to conduct Confirmatory Factor Analysis (CFA) and AMOS24.0, which is structural equation modeling software, was applied. The first fitting result of the model was poor - RMSEA was 0.096, which is significantly higher than 0.05, and related indexes such as NFI and CFI were lower than 0.9. The adjustment index showed that if the correlation among EOU, TG and TI was established and TI 3 parameters were removed, then the model's fitting degree of the model effectively improved. The results of confirmatory factor analysis after adjustment are shown in Fig. 2, and the fitting indexes are shown in Table 7.

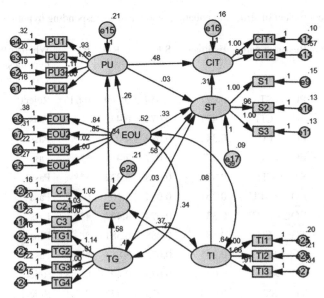

Fig. 2. Confirmatory factor analysis results

Table 7. Confirmatory factor analysis index

CHI/df	p-value	RMR	GFI	NFI	CFI	IFI	RMSEA
2.017	0.000	0.044	0.888	0.918	0.956	0.957	0.059

Table 7 confirms that NFI, CFI and IFI all exceeded 0.9, RMSEA is less than 0.06 and CHI/df falls below 3. Wen's (2004) proposed Chi square criterion standard confirms that this model's fitting index is acceptable.

Table 8 confirms all of the other hypotheses, with the exceptions of 1, 10 and 11, and the path coefficient, are also found to be significant at a 0.001 level (Hypothesis 6 is significant at the level of 0.01).

4 Discussion

Citizens' perceived usefulness and satisfaction have a significant positive impact on citizens' continuous-use intention to OGD. When the public perceive OGD to be more useful, they will be more satisfied with the data, and it is more likely that they will continue to use it. This study's hypotheses about continuous-use intention are all validated. This partially verifies Bhattacherjee's conclusion that user's continuance intention is affected by user's satisfaction and perceived usefulness, but perceived usefulness has a stronger influence than satisfaction. Citizens' experiences and attitudes towards OGD are closely related to their continuance intention to use OGD. Government departments should pay attention to the quality of OGD and improve the perceived usefulness of

Table 8. Path coefficient of structural equation model and corresponding hypothesis test results

Corresponding hypothesis	Research path	Estimate	S.E	C.R	P	Whether the hypothesis passes the verification
Hypothesis 1	PU - >ST	0.027	0.059	0.459	0.646	Reject
Hypothesis 2	PU - >CIT	0.477	0.074	6.442	***	Pass
Hypothesis 3	EC - >PU	0.342	0.084	4.071	***	Pass
Hypothesis 4	EC - >ST	0.577	0.077	7.45	***	Pass
Hypothesis 5	ST - >CIT	0.308	0.058	5.322	***	Pass
Hypothesis 6	EOU - >PU	0.261	0.084	3.103	0.002	Pass
Hypothesis 7	EOU - >ST	0.33	0.074	4.463	***	Pass
Hypothesis 8	TG - >EC	0.58	0.088	6.581	***	Pass
Hypothesis 9	TI - >EC	0.265	0.07	3.817	***	Pass
Hypothesis 10	TG - >ST	0.03	0.073	0.402	0.688	Reject
Hypothesis 11	TI - >ST	0.078	0.053	1.472	0.141	Reject

users, prioritize public demand when releasing government data and improve user satisfaction, as this will conceivably help to increase citizens' continuous-use intention to OGD.

In contrast *to previous IS research, citizens' satisfaction with OGD is affected by the degree of expectation confirmation and perceived ease of use, but not by perceived usefulness and their trust in government or network.* In addition to the expectation confirmation and perceived ease of use, the hypothesis related to factors that influence citizens' satisfaction with OGD has not been verified. Instead, expectation confirmation relates to the extent to which the citizens' perception of the performance of OGD corresponds to the initial expectation – in other words, with the extent to which the expectation is confirmed (see Table 1). This confirms that citizens' experience of OGD is of great importance. Government departments should concern themselves with constructing relevant platforms and the quality of open data, as this will help to improve citizens' perception of data. However, perceived usefulness does not significantly impact public satisfaction, this may be attributable to the citizens' affirmation of the usefulness of OGD.

In addition, citizens' *trust in the government and network has significant positive impact on their expectation confirmation.* Citizens' trust in the government has a dramatic and significant impact on expectation confirmation level; in addition, the results also show that public trust in the network has a relatively significant impact on expectation confirmation level. The greater the level of citizens' trust in their government, the more likely it is that they will continue to use OGD. The data analysis results support the Trust Model proposed by Belanger [18]. Government departments should work to improve credibility, as this will help to promote the sustainable development of OGD.

Due to the technical difficulties encountered by citizens when using OGD, perceived ease of use and expectation confirmation have significant positive effects on perceived usefulness. This verified research hypotheses related to the perceived usefulness of OGD. Public expectation confirmation and perceived ease of use will influence public perceptions about the usefulness of OGD. This confirms that citizens' understanding of data is subjective, and indicates that their demand and use experience will affect their attitude and behavior intention.

5 Conclusion

The research findings provide considerable insight into the current situation of OGD, which has a high open rate and low utilization rate, and also helped to identify existing problems. These findings will clearly interest relevant government departments as they seek to formulate appropriate data release policies. Future studies should endeavor to analyze the dynamic process through which citizens accept and utilize OGD.

References

1. Gasco-Hernandez, M., Martin, E.G., Reggi, L., Pyo, S., Luna-Reyes, L.F.: Promoting the use of open government data: cases of training and engagement. Gov. Inf. Q. **35**(2), 233–242 (2018)
2. Attard, J., Orlandi, F., Scerri, S., Auer, S.: A systematic review of open government data initiatives. Gov. Inf. Q. **32**(4), 399–418 (2015)
3. Gonzalez-Zapata, F., Heeks, R.: The multiple meanings of open government data: understanding different stakeholders and their perspectives. Gov. Inf. Q. **32**, 441–452 (2015)
4. Zeleti, A.F., Ojo, A., Curry, E.: Exploring the economic value of open government data. Gov. Inf. Q. **33**, 535–551 (2016)
5. Janssen, M., Charalabidis, Y., Zuiderwijk, A.: Benefits, adoption barriers and myths of open data and open government. Inf. Syst. Manag. **29**(4), 258–268 (2012)
6. Janssen, M., Konopnicki, D., Snowdon, J.L., Ojo, A.: Driving public sector innovation using big and open linked data (BOLD). Inf. Syst. Front. **19**(2), 189–195 (2017). https://doi.org/10.1007/s10796-017-9746-2
7. Zuiderwijk, A., Shinde, R., Janssen, M.: Investigating the attainment of open government data objectives: is there a mismatch between objectives and results? Int. Rev. Adm. Sci. Int. J. Comp. Public Adm. **85**(4), 645–672 (2018)
8. Ham, J., Koo, Y., Lee, J.N.: Provision and usage of open government data: strategic transformation paths. Ind. Manag. Data Syst. **119**(8), 1841–1858 (2019)
9. Zuiderwijk, A., Janssen, M., Dwivedi, Y.K.: Acceptance and use predictors of open data technologies: drawing upon the unified theory of acceptance and use of technology. Gov. Inf. Q. **32**(4), 429–440 (2015)
10. Talukder, M.S., Shen, L., Talukder, M.F.H., Bao, Y.: Determinants of user acceptance and use of open government data (OGD): an empirical investigation in Bangladesh. Technol. Soc. **56**(2), 147–156 (2019)
11. Lassinantti, J., Sthlbrst, A., Runardotter, M.: Relevant social groups for open data use and engagement. Gov. Inf. Q. **36**(1), 98–111 (2019)
12. Dawes, S.S., Vidiasova, L., Parkhimovich, O.: Planning and designing open government data programs: an ecosystem approach. Gov. Inf. Q. **33**(1), 15–27 (2016)
13. Duan, Y.Q., Qiu, X.T., He, S.Q.: Analysis on the status of China's urban government open data utilization from the thematic and regional perspectives. Libr. Inf. Serv. **62**(20), 65–76 (2018)

14. Zhu, H.C., Hu, X., Wang, X.B.: The user's continuance use intention of government data open platform based on the S-O-R framework. J. Mod. Inf. **38**(5), 100–105 (2018)
15. Gao, T.P., Mo, T.L.: Initial user adoption model and empirical study of government data open platform. E-Gov. **11**, 69–82 (2018)
16. Bhattacherjee, A.: Understanding information systems continuance: an expectation-confirmation model. MIS Q. **25**(3), 351–370 (2001)
17. Srivastava, S.C., Teo, T.S.H.: Citizen trust development for e-government adoption and usage: insights from young adults in Singapore. Commun. Assoc. Inf. Syst. **25**(1), 359–378 (2009)
18. Belanger, F., Carter, L.: Trust and risk in e-government adoption. J. Strat. Inf. Syst. **17**(2), 165–176 (2008)
19. Jiang, H., Wang, S., Shao, T.: Empirical analysis on factors influencing user acceptance of open government data. J. Xiamen Univ. Sci. Technol. **25**(4), 88–94 (2017)
20. Davis, F.D.: Perceived usefulness, perceived ease of use, and user acceptance of information technology. MIS Q. **13**(3), 319–339 (1989)

Industrial Control System Attack Detection Model Based on Bayesian Network and Timed Automata

Ye Sun[1], Gang Wang[1,2(✉)], Pei-zhi Yan[2], Li-fang Zhang[1], and Xu Yao[1]

[1] Mongolia Industrial University, Hohhot 010050, China
20191100107@imut.edu.cn
[2] Information Construction and Management Center, Hohhot, China

Abstract. The current industrial control system attack detection methods are single, the detection results are fuzzy and cannot be applied to the domestic industrial environment. In response to the above problems, an industrial control system attack detection model based on Bayesian network (BN) and Timed automata (TA) theory is proposed. First, collect the real industrial purification data of the aluminum factory, that is, the sensor and actuator signals, and preprocess the signals through time compression, segmentation, and queue division; secondly, establish Timed automata and Bayesian network models respectively, using probability time automatization The computer simulates the regular behavior of the time series, and at the same time uses the Bayesian network to build the dependency relationship between the sensor and the actuator; finally, the model's detection result of the attack data is calculated. Theoretical analysis and experimental results show that compared with Deep Neural Network (DNN) and Support Vector Machine (SVM), the model in the article has improved time and accuracy.

Keywords: Bayesian network · Timed automata · Industrial control system · Attack detection

1 Introduction

Industrial control system (ICS, industrial control system) is an important part of the country's critical infrastructure. The number of attacks it suffers is increasing year by year [1]. Although the number of attacks is much lower than that of Internet attacks, every time Compared with Internet attacks, attacks will have a huge impact on the country's industrial production, and the economy will suffer heavy losses [2]. At the beginning of the 20th century, the United States began to pay attention to the safety of industrial control systems, and established number of national-level laboratories, established critical infrastructure test ranges, and promoted the implementation of two national-level special programs in 2012, namely, the national SCADA test bed. The Plan (NSTB) and the Control System Security Plan (CSSP) provide a solid foundation for the development of industrial control system security research. Although my country has not yet seen a large-scale industrial control system security problem, it is still in the basic

© Springer Nature Switzerland AG 2022
J. Wei and L.-J. Zhang (Eds.): BigData 2021, LNCS 12988, pp. 79–92, 2022.
https://doi.org/10.1007/978-3-030-96282-1_6

research stage in terms of industrial control system security research. There are still up to 91% of industrial control systems in my country that use foreign brands, and there is a great infection of industrial viruses. Potential hidden dangers [3]. Therefore, it is very important to take corresponding protective measures for attack detection in industrial control systems. At present, the various industrial control system attack detection methods proposed are mainly through the acquisition of industrial control system monitoring and data acquisition (SCADA) workstations, human machine interfaces (HMI), programmable logic controllers (plc), and low-level communications. Network traffic uses methods such as signature-based [4, 5], verification [6, 7], behavioral norms [8], and machine learning [9, 10]. Signature-based attack detection methods require an up-todate signature dictionary of all known attacks. As the number of unknown threats continues to increase, this becomes increasingly infeasible; verification methods basically use formal models to test a certain attack at the source code level. Whether these signals have large deviations from the values specified in the industrial control system design. Although powerful, complete verification based on source code is usually not feasible due to the state explosion problem. In the state explosion problem, the resulting model becomes too large to be analyzed; the behavior-based method requires an accurate understanding of the industrial control system. Behavior, obtaining such knowledge may require certain expert experience; attack detection methods based on machine learning, such as deep neural networks (DNN) and support vector machines (SVM) [11] mostly focus on detecting anomalies in the feature space of industrial control systems that find data points that deviate from the normal space. This requires little system knowledge and can detect a wide range of attacks, but the disadvantage is that there is less understanding of the industrial process of industrial control systems, and the algorithm is more complex. When faced with large-scale data, it takes a long time.

Aiming at the shortcomings of machine learning, this article provides an industrial control system attack detection method based on Bayesian networks and timed automata. First, it explains the data set used and analyzes the specific industrial control system-the process flow of the electrolytic aluminum plant. And attack scenarios; secondly, use Bayesian network and timed automata to build a graph model to detect industrial control system attacks, give corresponding attack detection results and compare related methods currently used; finally, research on industrial control system attack detection Carry out prospects and put forward the research and development direction of related issues.

2 The Security of Electrolytic Aluminum Purification (SEAP) Data Set

The Seap data set used in this article was obtained from the flue gas purification process of an aluminum factory. The data set includes normal working data and attack data. The flue gas purification process (see Fig. 1), and the important sensor and actuator parameters in the figure are described in Table 1.

The attack data is set in the electrolytic aluminum purification system based on the attack scenarios given by the industrial control system attack model. A generalized model for cyber-physical systems with the attacker's intention space [12] (see Fig. 2) According to the attack model, the attacks based on the two scenarios used in this paper

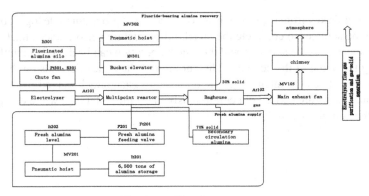

Fig. 1. Purification process

are given [13]. The first is to attack basic components and main attributes of the industrial control system, the attack point is located as a sensor, and the data operation; the second is to industrial control.

Table 1. Purification process parameters

Number	Stage	Parameter	Meaning
1	1	At101	Inlet hydrogen fluoride concentration
2	1	MV101	Whether the multi-point reactor is running
3	1	MV102	Whether the bag filter is running
4	1	MV103	Whether the bag filter blows back
5	1	MV104	Blowing cycle of bag filter
6	1	At102	Dust concentration at outlet of bag filter
7	1	MV105	Whether the main exhaust fan is running
8	2	It201	6,500 tons of total warehouse material level
9	2	It202	Fresh alumina bin material level
10	2	P201	Opening of fresh alumina feeding valve
11	2	Ft201	Fresh alumina cutting flow
12	2	MV201	Whether the pneumatic hoist is running
13	3	It301	Fluorine-loaded alumina bin material level
14	3	MV301	Whether the pneumatic hoist is running
15	3	MV302	Whether the bucket elevator is running
16	3	Pt301	Chute fan pressure
17	3	S301	Chute fan frequency

The system's key from the sensor is transmitted to the programmable logic controller to control the actuator to complete the attack infrastructure performance attack, that is, the attack location is the actuator, and the attack operation can be directly completed by modifying the value of the actuator. The duration of the attack depends on the type of attack and the attacker's intentions. The duration of each attack ranged from 101 s to 10 h. Some attacks are carried out continuously in 10-min intervals, while others allow time for the system to stabilize.

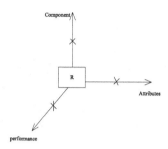

Fig. 2. Attack model

Given an attack scenario (see Fig. 3) a), the final attack result is shown in Figure b). That is, attacking the fresh alumina feeding flow rate Ft201, causing the fresh alumina feeding valve P201 to close, the fresh alumina supply is suspended, and the amount of fresh alumina in the multi-point reactor is gradually reduced, until there is almost no, the

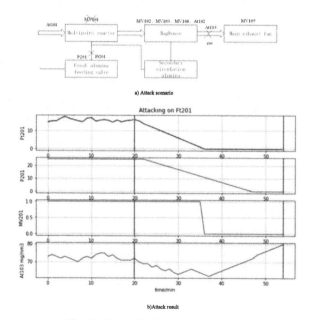

Fig. 3. Example of an attack scenario

multi-point reactor is suspended, namely MV101 The value changes from 1 to 0, and the multi-point reactor is used as a reaction vessel for alumina and fluorinecontaining flue gas. When the reaction is suspended, the final outlet hydrogen fluoride concentration increases, that is, the At103 value increases. Exhaust will cause environmental pollution of toxic gases.

3 Signal Processing

This section discusses the preprocessing process of processing high-dimensional noise signals in the purification system. In Table 1, sensors and actuators are divided according to each stage. For sensors that measure concentration, pressures and frequency, as well as the manually operated 6500-ton total bin material level, there are large noises and subtle trends, and only thresholds are used for detection. Because of their relatively obvious regular patterns and their dependence on actuators at the same stage, the material level of the fresh alumina silo and the fluorine containing alumina silo can learn their sequence behavior through timed automata and Bayesian networks.

3.1 Time Compression

The sensor signal uses an averaging filter for time compression at each initial stage. Since the original signal of the collected data set is transmitted every 10 s, the time span of the data is not obvious, and it is difficult to observe the law. For this reason, the signal needs to be time compressed. The following is the time compression process of this signal. The definition (1) of the one-dimensional time series of the sensor signal is:

$$x[n] = [x_1, x_2, ..., x_n] \tag{1}$$

This article uses a naive average filter to compress the time series, the (2) is:

$$\bar{x}[w] = [\bar{x}_1, \bar{x}_2, ..., \bar{x}_w] \tag{2}$$

The calculation (3) for the i-th element of is:

$$\bar{x}_i = \frac{w}{n} \sum_{j=n/w(i-1)+1}^{n/w*i} x_j \tag{3}$$

For simplicity and clarity, define n/w = 6, assuming that n is divisible by w. If not, it can be modified by loading and adding to the remainder of x to average. The original signal of the fluorine-containing alumina bin material level and the time-compressed signal (see Fig. 4).

3.2 Signal Segmentation

Piecewise linear representation (PLR), as one of the most commonly used preprocessing methods for time series segmentation, has been used by many researchers to support clustering, classification, indexing and association rule mining of time series data [14].

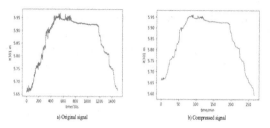

a) Original signal   b) Compressed signal

Fig. 4. Time compression of fluorine-containing alumina bin material level

In this paper, the sliding window differential division (SWIDE) algorithm is used to perform piecewise linear approximation to the sensor signal. The working principle of the SWIDE algorithm is to anchor the left point of a potential segment on the first data point of the time series, and then try to approach the right point of the data by adding a longer segment. At a certain point i, if the error of the potential segment is greater than the threshold specified by the user, the subsequence from the anchor point to i − 1 is converted into a segment. Then move the anchor point to position i and repeat this process until the entire time series is converted into segments. The pseudo code of SWIDE is shown in Algorithm 1 [14]. Finally, the segmented sensor signal (see Fig. 5).

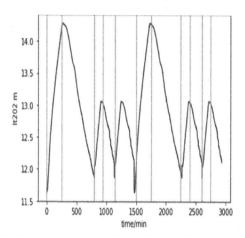

Fig. 5. Segmented signal of fresh alumina bin material level

3.3 Time Alignment

The clustering algorithm is used to discretize and symbolize the sensor signal, so that the sensor and actuator signal segments can be time aligned, and the timed automata and Bayesian network are used for learning. The material level of the fresh alumina silo in Fig. 6 shows three discrete signals described in the alphabet. The curve trends of a large increase (a), a small increase (b) and a decrease (c) are obvious and have specific physical meanings. which is:

(a) The curve rises sharply. At this time, the amount of feed for one day of manual operation enters the fresh alumina warehouse, and the material can be used for about 6–7 h.

(b) The curve rises slightly. At this time, the circulating alumina enters the reactor, and a small amount of fresh alumina is manually added.

(c) The curve drops, fresh alumina is delivered to the reactor, and the descending amplitude is controlled by the feeding valve.

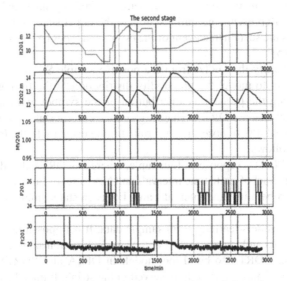

Fig. 6. Time alignment based on segmentation and clustering results

From this, the time string of the alignment sensor of the sensor and actuator in the second stage (see Fig. 6) is: (a, 250) (c, 550) (b, 150) (c, 200) (b, 150) (c, 210)), (a, 240) (c, 560) (b, 140) (c, 210) (b, 130) (c, 230). The misalignment of the misaligned segment may be due to noise in the original signal.

4 Fusion Model

This section will introduce timed automata and Bayesian networks and their learning algorithms. The input of the timed automata learning process is composed of the time string obtained by the queue, and the output can represent the normal working behavior of the sensor in the industrial control system. The input of the Bayesian network learning process is composed of time-aligned data of sensors and actuators in a queue, and the output can represent the relationship between actuators and sensors in an industrial control system.

4.1 Probabilistic Timed Automata

This article will first introduce the common real-timed automata (DRTA) in the comparative research, and then introduce the PDFA probabilistic deterministic real-timed automata (PDRTA), which is the model used in this research. The discrete events of industrial control system sensors in this article are represented by timing strings, where ai is a discrete event that occurs with a time delay from the $i - 1$th event, and is the duration of each event.

Definition 1: DRTA Let RTA be a five-tuple, where Q is a finite state set, is a finite event set, is a finite transition function state set, is the starting state, and is a finite set of acceptable states, Represents the initial state and the target state, represents one of the events, represents the sustainable event of the event, and the RTA that contains only one clock is DRTA [15].

In order to identify the DRTA from the positive sample S+, it is necessary to use the DRTA structure to model the probability distribution of the timing string. This is achieved by adding the probability distribution of the symbol and time value of the timed event to each state of DRTA. That is, given the current state q of PDRTA, that is, represents the probability of observing a certain timing event (a, t), the formal definition of PDRTA is as follows.

Definition 2: PDRTA supposes a four-tuple, where is a DRTA that does not contain the final state, H is a set of finite time intervals and N, is a set of finite probability events, is a set of finite probability time intervals, and. For each state, among them. It is called a PDRTA [15].

DRTA only considers learning from sample data. It is impossible to judge whether there are normal data and abnormal data in the attack detection sample data. Therefore, this paper chooses PDRTA to build the model. In DRTA and PDRTA, the state is a latent variable that cannot be directly observed in the string. It must be estimated using a learning method. Among them, the machine learning algorithm RTI+ is used to learn human behavior from unlabeled data [16, 17], is the key algorithm used in the PDRTA learning process.

Compared with the traditional state machine learning algorithm, the RTI+ algorithm [17] constructs a large tree shaped automaton from the input string samples, namely the augmented prefix tree (APTA), which uses the timed APTA. Each state of the tree can be realized by an unnamed string, so the input samples can be accurately coded. For timed automata learning, the initial values of the lower and upper limits of all durations are set to the minimum time value and the maximum time value from the input sample s. State merging and splitting are the two main operations of structure and time learning in RTI+. The algorithm will greedily merge the state pairs (q, q') in this tree to form a smaller and smaller tree.

Finally, the PDRTA model of the material level of the fresh alumina silo is obtained. After the first manual feeding of the fresh alumina silo, the opening of the feeding valve is relatively large, so it takes about 13 h to complete, and then the alumina is recycled. Do a small amount of manual feeding at intervals, and after about 2 cycles, start to re-feed on the next day to cycle.

4.2 Bayesian Network

Bayesian network, also known as belief network, or directed acyclic graph model, is a probabilistic graph model, first proposed by Judea Pearl [18] in 1985. It is an uncertainty processing model that simulates the causal relationship in the human reasoning process, and the network structure is a directed acyclic graph (DAG). This article describes the dependency between actuators and sensors through Bayesian networks. The nodes in the directed acyclic graph of Bayesian networks represent random variables, which can be observed variables or hidden variables, unknown parameters, etc. The arrow connecting the two nodes represents whether the two random variables are causally related or unconditionally independent. Let graph G = I, E represents a directed acyclic graph (DAG), where I represents the set of all nodes in the graph, and E represents the set of connected line segments, and let it be a certain of the directed acyclic graph The random variable represented by the node i, if the joint probability of the node can be expressed as (4):

$$P(x) = \prod p(x_i | xPa(i))i \in I \tag{4}$$

Then call X a Bayesian network relative to a directed acyclic graph G, where Pa(i) represents the parent of node i. In addition, for any random variable, its joint probability can be obtained by multiplying the respective local conditional probability distributions to obtain (5):

$$p(x_1, x_2, ..., x_k) = p(x_k | x_1, x_2, ..., x_{k-1})...p(x_2 | x_1,)p(x_1) \tag{5}$$

Bayesian network learning includes structure learning and parameter learning. This paper uses the greedy search algorithm K2 [19] for structure learning. The general idea is as follows. Initially each node has no parent node. Then, it incrementally adds the parent item that causes the score of the resulting structure to increase the most. When adding any single parent node cannot increase the score, it will stop adding parent nodes to the node. The random variable of the sensor is fixed as the last item, assuming that it is not the parent node of any other variable, and the order of the parent nodes is random. Parameter learning is relatively simple. When learning the structure, maximum likelihood estimation is used, that is, the probability of each node is calculated from the data to obtain the conditional probability (CPD) table.

The API interface of HUGIN is called to obtain the dependence of sensors and actuators in the fresh alumina supply (see Fig. 7), the Bayesian network is composed of graph structure and parameters. The parameter is expressed by a conditional probability distribution, which summarizes the probability distribution of a given parent node.

5 Experimental Evaluation

5.1 Experimental Design

The implementation framework of this method is scikit learn and scipy, the operating system is windows10, the processor is Intel (R) Core (TM) i5-10210U CPU with 8 cores,

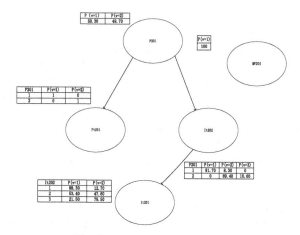

Fig. 7. Bayesian network model

and the memory is 16 GB. In this experiment, DNN uses both literature and SVM Using the default parameters in the literature [10], the method in this paper uses the timed automata training model and the Bayesian network model generated by the HUGIN tool to be fused as the last result for experimental comparison and verification.

The data sources mainly come from domestic real data. The evaluation of industrial control system intrusion detection methods mostly uses foreign data sets. Although the current foreign data is relatively mature and widely used, the data does not meet the key domestic industrial control foundations. Facility data cannot be truly used in domestic industrial control systems. Based on the above situation, this paper uses the Seap data set as the evaluation data, which is generated by the flue gas purification system of a domestic aluminum factory. The data set includes 18 key infrastructure data and 1 sign data, which involves 8 sensors and 10 actuators. At the same time, according to the real industrial control environment attack scenario, this article selects most of the normal data and a small part of the attack data to construct a balanced data set. 10800 pieces of data are selected, and the normal data and the attack data are processed according to 8:2. Divide, and then randomly select 70% of the data as the training set and 30% of the data as the test set.

5.2 Evaluation Indicators

Redefinition of TP and CP

Traditional anomaly detection methods are based on data points. Each piece of time series data is regarded as an instance, which cannot highlight the continuity of time series data in time. Common industrial control system attacks usually last from a few minutes to a few hours. This article uses the anomaly detection method for time series data to redefine true cases (TP) and false positive cases (FP). As shown in Fig. 8, FP refers to the difference between the real time sub-sequence data and the detected sequence data.

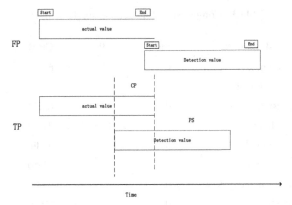

Fig. 8. Redefinition of TP and CP

Overlapping (see Fig. 8). For TP, it is divided into coverage percentage (CP) and non-covered data volume (PS) Among them, the meanings of TP, FP, FN, and TN are shown in Table 2.

Table 2. Meaning of TP, FP, FN, TN

The true situation	Predict	
	Normal	Attack
Normal	TP	FN
Attack	FP	TN

Calculation of Accuracy, Recall Rate and F-measure
The common comparison indexes of machine learning algorithms are accuracy, recall and F-measure.

Time Cost
Time overhead is an important indicator to measure the performance of artificial intelligence models. This paper uses the training time and test time of the computing algorithm as a measure of the time overhead performance of the model. By comparing the time used by the deep neural network (DNN), support vector machine (SVM) and the fusion model, it shows the effectiveness of the algorithm application in the real environment.

5.3 Experimental Results
Through the above three performance indicators, at the same time, because some signals do not have time rules, the threshold value is used to judge them here, and the threshold value range is the maximum and minimum value of the signal value. Experiments on the fusion model in this paper. The analysis of the experimental results is as follows:

Table 3. Comparison of model and single method

Stage	method	FP	TP	CP(%)	PS
1	Fusion model	0	10	63.50	22
	Time automata	无	无	无	无
	Bayesian network	0	23	**83.30**	10
	Threshold	0	15	71.50	17
2	Fusion model	0	11	**90.00**	6
	Time automata	3	20	51.50	29
	Bayesian network	0	16	65.00	21
	Threshold	0	28	50.50	30
3	Fusion model	0	30	**82.00**	11
	Time automata	5	22	47.50	31
	Bayesian network	0	18	60.05	24
	Threshold	0	11	59.40	24

Table 4. Attack scene detection

Stage	Attack scenario	FP	TP	CP(%)	PS
1	1	3	20	**87.50**	29
	2	4	15	77.50	34
2	1	0	30	55.00	135
	2	2	23	**64.50**	83
3	1	3	27	**70.00**	81
	2	0	32	62.00	128

According to the redefinition of TP and CP, the fusion model in this paper is compared with a single method. The experimental results are see Table 3 and Table 4.

At the same time, the accuracy, recall and F-measure comparison results of the deep neural network (DNN) and support vector (SVM) methods used in this article and the existing literature are shown in Table 5.

It can be observed that the fusion model in this paper achieves a better balance than DNN and SVM in terms of accuracy and recall. Moreover, because the model in this paper does not need to perform feature selection, it is relatively simple in terms of algorithm complexity compared to SVM and DNN. The experimental results in running time are shown in Table 6. The time required for industrial control system inspection based on the fusion model is greatly reduced.

The experimental results show that the method in this paper successfully identifies attacks against industrial control systems. Compared with DNN and SVM, its F-measure

Table 5. DNN, SVM and fusion model results

Method	Accuracy	Recall rate	F-measure
DNN	**0.92300**	0.68477	0.78623
SVM	0.91525	0.67901	0.77962
Fusion model	0.92161	**0.79987**	**0.85644**

Table 6. Time comparison

Method	Training time	Test time
DNN	410 s	135 s
SVM	355 s	100 s
Fusion model	**214 s**	**56 s**

value is larger, the performance is more balanced, and the time used is shorter. Among them, the SVM algorithm has the worst accuracy and takes a long time. The reason is that the accuracy of the algorithm depends on the selected kernel function, but there is no good way to solve the selection of the kernel function; the second is when facing large-scale training Samples, the kernel matrix that needs to be accessed, will consume a lot of time and space. The training time used by the DNN algorithm is longer because the algorithm itself has a complex structure, and the selection of the activation function and the initialization of the weights will affect the training time of the DNN algorithm. The model in this paper may not be as accurate as DNN. This is because the model is based on industrial processes, and there may be process incompleteness, and the signal data is noisy, but its equalization performance is better than DNN and SVM algorithms. By calling API The Bayesian network and timed automata model algorithm formed are simpler than DNN, so the required time is shorter.

6 Conclusion

In this paper, Bayesian network and timed automata are used to study the sensor and actuator signals in industrial control system attack detection based on the time-dependent characteristics of time series. The corresponding attack scenarios are rationally designed and the fusion model is further established. Finally, the validity and practicability of the model in this paper are verified. At the same time, this article also has problems such as signal noise processing and time definition of attack scenarios. The improvement and processing of these aspects will be the next research content.

Fund Project:. Fund of Education Department of Inner Mongolia Autonomous Region [NJZZ18077]

References

1. Wenli, S., Panfeng, A., Ming, W.: Overview of research and development of intrusion detection technology for industrial control systems. Appl. Res. Comput. **34**(002), 328–333 (2017)
2. Yingxu, L., Zenghui, L., Xiaotian, C.: Overview of industrial control system intrusion detection research. J. Commun. **38**(002), 143–156 (2017)
3. Yue, W., Ting, F., Minhu, M.: The evolution and enlightenment of U.S. critical infrastructure information security monitoring and early warning mechanism. Intell. Mag. **32**(002), 142–155 (2016)
4. Morrris, T., Wei, G.: On cyber attacks and signature based intrusion detection for MODBUS based industrial control systems. J. Digit. Forensics Secur. Law. **9**, 37–56 (2014)
5. Oman, P., Phillips, M.: Intrusion detection and event monitoring in SCADA networks. DBLP. **253**, 161–173 (2007)
6. Zheng, X., Julien, C.: Verification and validation in cyber physical systems. In: Research Challenges and a Way Forward. IEEE (2015)
7. Zuliani, P.: Statistical model checking for cyberphysical systems. In: Proceedings of the International Conference on Automated Technology for Verification and Analysis (2011)
8. Adepu, S., Kang, E., Jackson, D., et al.: Model-based security analysis of a water treatment system. In: Proceedings of the 2nd International Workshop on Software Engineering for Smart CyberPhysical Systems (2016)
9. Junejo, K.N., Goh, J.: Behavior-based attack detection and classification in cyber physical systems using machine learning. In: Proceedings of the ACM International Workshop on Cyber-physical System Security (2016)
10. Goh, J., Adepu, S., Tan, M., et al.: Anomaly detection in cyber physical systems using recurrent neural networks. In: Proceedings of the IEEE International Symposium on High Assurance Systems Engineering (2017)
11. Inoue, J., Yamagata, Y., Chen, Y., et al.: Anomaly detection for a water treatment system using unsupervised machine learning. In: Proceedings of the 2017 IEEE International Conference on Data Mining Workshops (ICDMW) (2017)
12. Adepu, S., Mathur, A.: Generalized attacker and attack models for cyber physical systems. In: Proceedings of the Computer Software and Applications Conference (2016)
13. Adepu, S., Mathur, A.: Using process invariants to detect cyber attacks on a water treatment system. In: Proceedings of the 31st International Conference on ICT Systems Security and Privacy Protection - IFIP SEC 2016 (2016)
14. Keogh, E., Chu, S., Hart, D., et al.: An online algorithm for segmenting time series. In: Proceedings of the 2001 IEEE International Conference on Data Mining (2002)
15. Verwer, Ewout, S.: Lazy abstraction for timed automata: theory and practice. Electr. Eng. Math. Comput. Sci. (2013)
16. Herbreteau, Walukiewicz, l.: Efficient verification of timed Automata. J. Comput. Sci. **07**(004), (2016)
17. Verwer, S., de Weerdt, M., Witteveen, C.: A likelihood-ratio test for identifying probabilistic deterministic real-time automata from positive data. In: Sempere, J.M., García, P. (eds.) Grammatical Inference: Theoretical Results and Applications, vol. 6339, pp. 203–216. Springer, Heidelberg (2010). https://doi.org/10.1007/978-3-642-15488-1_17
18. Cooper, G.F., Herskovits, E.A.: Bayesian method for the induction of probabilistic networks from data. Mach. Learn. **9**(4), 309–347 (1992)
19. Bielza Lozoya, M.C., Moral Callejón, S., Salmerón Cerdán, A.: Recent advances in probabilistic graphical models. Int. J. Intell. Syst. **30**(3), 207–208 (2015)

Design on Flood Control and Disaster Reduction System for Emergency Management Based on BDS

Siyao Wang[1], Hao Feng[1(✉)], Jing Wang[1], Xuejiao Peng[1], Yumeng Zheng[1], Xiaohu Fan[1,2], and Quan Yuan[1]

[1] Department of Information Engineering, Wuhan Collage, Wuhan 430212, China
{19202130107,19202030113,19202130109}@mail.whxy.edu.cn, {8206, 8335,9093,9420}@whxy.edu.cn
[2] PIESAT International Information Technology Co. Ltd., Beijing, China

Abstract. In order to deal with the heavy social losses caused by the heavy rainfall in Shanxi Province, China, the flood control and disaster reduction system based on the BDS (BeiDou Navigation Satellite System) is designed to integrate multiple subsystems such as disaster reporting, disaster warning and disaster communication, which can collect and analyze the data information of the affected areas, the location of the affected people, the number of casualties, disaster losses and other aspects, using massive satellite remote sensing and sensor data to do analysis, to achieve accurate control of the flood situation. Before the disaster occurs, the areas that may occur disasters are fitted by UAV aerial survey technology and DEM (Digital Elevation Model), and disaster warning is issued in advance. In the event of a disaster, the BeiDou short message service is used to locate and search the affected people. After the occurrence of disasters, we can build databases to share disaster information and summarize disaster relief experience. So as to maximize the protection of people's life and property safety, reduce social and economic losses.

Keywords: BeiDou Navigation Satellite System · Emergency management · Remote sensing · K8S

1 Introduction

The terrain of Shanxi Province is complex, mainly a large area of loess covered mountain plateau, the terrain is high in the northeast and low in the southwest. The interior of the plateau is undulating and the water vapor in the valley is easily condensed. By this year in the western Pacific subtropical high and the influence of the westerlies moisture is sufficient, in lu-liang mountains, taihang water vapor by forming strong rainfall, topography uplift of the loess soil loose after rain [1], rivers and so on to carry large amounts of sediment [2], by the surrounding hills together to the middle plain resulting in huge casualties and property losses.

© Springer Nature Switzerland AG 2022
J. Wei and L.-J. Zhang (Eds.): BigData 2021, LNCS 12988, pp. 93–106, 2022.
https://doi.org/10.1007/978-3-030-96282-1_7

2 Related Works

2.1 Emergency Management Systems

Typical cases have been accumulated in the development of emergency management systems for natural disasters and accidents. Developed countries and regions such as the United States, Europe and Japan are actively developing emergency management systems, shown in Table 1.

Table 1. Cases of emergency management system

Country/Organization	Appropriate authority	Typical case	Reference website
The United Nations	Caribbean Catastrophe Risk Insurance Fund (CCRIF)	Develop and strengthen the information base of key natural disaster risk; Regional studies on the economics of climate change and the impact of natural disasters on specific sectors such as tourism;	Caribbean Catastrophe Risk Insurance Facility (CCRIF) - CARICOM
	Economic Commission for Latin America and the Caribbean (UNECLAC)	Develop decision-making tools to help mitigate the effects of natural disasters; Improve climate data and information systems to support emergency planning and achieve broader disaster resilience;	Economic Commission for Latin America and the Caribbean (cepal.org)
	Unu Institute for Environment and Human Security (UNU-EHS)	A new assessment tool for low - and middle-income countries that compares climate and disaster risk with short-term resilience;	Institute for Environment and Human Security (unu.edu)
America	Federal Emergency Management Agency	Seismic Rehabilitation Cost Estimator	Federal Emergency Management Agency (FEMA)
		Disaster assessment management system loss estimation software	
	United States Geological Survey	American National Earthquake System	USGS.gov I Science for a changing world
	National Science Foundation	Natural disaster engineering research infrastructure	NSF - National Science Foundation

(*continued*)

Table 1. (*continued*)

Country/Organization	Appropriate authority	Typical case	Reference website
Japan	East Japan Railway	Earthquake Early Warning System for the Shinkansen	JR-EAST - East Japan Railway Company (jreast.co.jp)

In terms of meteorological disaster prediction [4], a comprehensive typhoon monitoring system based on meteorological satellites [5]. Doppler weather radar and automatic weather stations has been basically completed, and its capability has reached the advanced level in the world. The built typhoon service platform covers functional modules such as typhoon track display and retrieval, forecast making and early warning information release. For the National Meteorological Center [6], East China Regional Meteorological Center, South China Regional Meteorological Center and other business units to provide a wealth of products. China has also built the first-generation national emergency response platform [7], established a cross-field, cross-level, cross-time and cross-regional "one map for emergency response" collaborative consultation system [8], and developed a on-site 3D detection system and mobile emergency response platform equipment for the "last ten kilometers" information acquisition problem in the emergency scene [9].

Compared with the international level, although the development of science and technology the wisdom in China is better, but some of the key instruments and equipment import dependence degree is higher, the development is still in a passive position, key data such as land surveying and mapping data, the geological survey data, disaster census data respectively held by different functional departments, such as lack of timeliness, sharing data;Lack of computational and analytical tools for large-scale natural disasters.

In the face of major natural disasters, it is necessary to complete a series of work in a short time, such as information acquisition, information collection, data analysis, solution formation, personnel and materials allocation and rescue, which involves information integration, big data processing, short message service, high-precision positioning and other elements. Then, the disaster relief command subsystem is used to obtain the distribution information of personnel and materials and relevant rescue routes, and the information sharing platform is used to specify the most suitable disaster relief plan for the disaster area in a short time by calling the key information such as land mapping data and geological survey data, so as to control the disaster situation in the shortest time.

2.2 Navigation Systems

The world's four largest satellite navigation systems are the GLOBAL Positioning System (GPS) of the United States, Russia's GLONASS, the European Union's GALILEO and China's Beidou Navigation Satellite System (BDS) [10]. The United States is gradually upgrading the SPACE segment and the ground control end of GPS. In the space segment, the interoperable signal L1C in L1 band is added. The design life is increased to 15 years, the positioning accuracy is 3 times of the original model, and the anti-interference ability is 8 times of the original. At the same time, the next-generation control system OCX is combined with GPS-III satellite on the ground to improve the accuracy of satellite navigation system [11]. Russia expects to deploy six satellites in the high orbit segment of the system by 2025. The first satellite will be operational in 2023, which will improve the navigation accuracy of GLONASS by 25% over half the earth. Galileo Satellite Navigation System By 2020, the planned 30 satellites have all been deployed, consisting of 24 service satellites and 6 in-orbit standby satellites, located in the three circular earth orbit planes (MEO) 23,222 km above the Earth [12].

2.3 Introduction of Each System

GPS system, with its characteristics such as high precision, all-weather, high efficiency and multi-function has been widely used, but as the big dipper, the continuous development of Galileo satellite navigation system, GPS system increasingly shows the inherent defect of many aspects such as the signal strength is weak, difficult to penetrate the building block, the public, are susceptible to interference and navigation message signal must be updated once a day, It is difficult to guarantee the reliability of navigation service (Table 2).

Table 2. GPS system parameters

Satellite model	Service time	Signal types	point/MHz	Life/years	Add features
GPS-IIR	1997–2004		1 227.60	8.5	Board clock detection function
GPS-IIR(M)	2005–2009	1.2(C)		7.5	M-code anti-interference
GPS-IIF	2010–2016	1.5	1 227.60	12.0	Advanced atomic clock, enhanced signal strength and quality
GPS-III	2016-	1.1(C)	1 176.45		No SA, enhancing signal reliability
GPS-IIIF	2018-	1.1(C)	1 176.45	15.0	Laser reflection function, load search

Russia sent into orbit the new generation 15 GlONASS-K satellite, which carries two types of navigation signals - frequency separation signals and code separation signals. At the same time, GLONASS system not only retains frequency division multiple access (FDMA) signal but also introduces code division multiple access (CDMA) signal and improves the stability of spaceborne clock. However, GLONASS has a large orbital inclination Angle, which is not suitable for low latitude areas, and there is a disadvantage, which is related to the fact that Russia is located in high latitude areas (Table 3).

GALILEO can operate effectively and work well. As a part of the existing INTER-NATIONAL Satellite-assisted Search and Rescue Organization satellites, GALILEO

Table 3. GLONASS system parameters

GLONASS satellite	Service time	FDMA/MHz		CDMA/MHz		
		$1\,602 + n \times 0.562\,5$	$1\,246 + n \times 0.437\,5$	$1\,600.995$	$1\,248.06$	$1\,202\,25$
GLONASS-M	2013–2016	L1OF, L1SF	L2OF, L2SF			L3OC^2
GLONASS-K	2011/2018	L1OF, L1SF	L2OF, L2SF			L3OC
GLONASS-K2	2017-	L1OF, L1SF	L2OF, L2SF	L1OC, L1SC	L2OC, L2SC	L3OC

can provide positioning accuracy within 2000 m for 77% of rescue positions and within 5000 m for 95% of rescue positions. The quality of Galileo observations is good, but the stability of the satellite clock is slightly poor, resulting in poor system reliability.

The Beidou Navigation Satellite System (BDS) is a global navigation satellite system [13] independently built and operated by China with the needs of national security and economic and social development in mind. It is an important national space-time infrastructure that provides all-weather, round-the-clock and high-precision positioning, navigation and timing services to users around the world. BDS can provide bidirectional high-precision timing and short message communication services [14]. Its position accuracy is plane 5, elevation 10 m, speed measurement accuracy 0.2 m/s, and timing accuracy 50 ns.

2.4 Technical Comparative Analysis (See Table 4)

GPS uses 6-track plane, but GLONASS, Galileo and BDS all use 3-track plane, which simplifies the maintenance and configuration of constellation network, reduces resource consumption and network operation and maintenance. Moreover, the geometric distribution of 3-track satellites is higher than that of 6-track plane. Compared with THE hybrid constellation of MEO and GEO satellites used by GPS, BDS uses the hybrid constellation composed of MEO, IGSO and GEO satellites, which can effectively improve the accuracy of user positioning and improve the performance of the system, especially in terms of system stability and availability. Therefore, this paper chooses the BDS system with more mature technology as the carrying platform of flood control and emergency response system.

3 Top-Level Design

3.1 Framework

For the overall design of the flood control and disaster reduction system, the branch structure design method can be used to split the flood control emergency system from the large frame into several relatively independent modules. The framework diagram of

Table 4. GPS/BDS positioning accuracy statistics

Positioning system	Positioning accuracy in all directions														
	DGPS/m			Static baseline/cm			PPP/cm			RTK/cm			Network RTK/cm		
	N	E	U	N	E	U	N	E	U	N	E	U	N	E	U
GPS	0.26	0.25	0.37	0.38	0.09	0.06	0.82	0.73	1.62	1.43	0.85	2.01	1.26	1.32	2.78
BDS	0.33	0.31	0.52	0.73	1.14	0.43	0.96	0.83	2.32	1.45	0.87	1.98	1.71	1.65	3.45
GPS/BDS	0.20	0.18	0.26	0.44	0.14	0.45				0.89	0.78	1.83	1.22	1.13	3.23

the flood control emergency system based on BDS is listed based on the relevant system objectives mentioned above, shown in Fig. 1.

Fig. 1. Functional structure of flood control and disaster reduction system

3.2 Architecture

The flood control and emergency system is mainly divided into four layers: data layer, service layer, application layer and user layer. The data layer mainly provides information support for the system platform. The service layer mainly provides interface support and technical support. The application layer distributes different functions of each subsystem, which is the function embodiment of flood control and emergency system. The user layer is mainly used by different levels of users, as shown in Fig. 2.

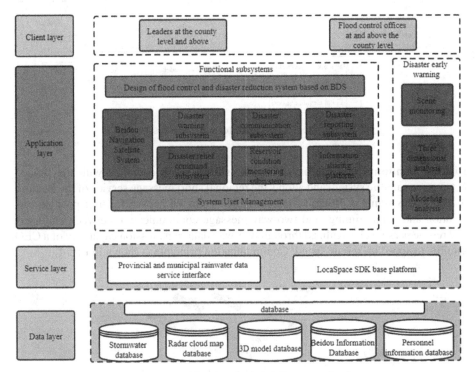

Fig. 2. Architecture of flood control and disaster reduction system

3.3 Modules and Subsystems

Disaster Warning Subsystem. Superposition of the subsystem using the UAV aerial technology, irregular triangulation TIN algorithm fitting geographic region contour map and the 3 d information, implementation of the construction of the digital elevation model (DEM, the region integrated radar chart, meteorological satellite cloud pictures, such as meteorological hydrological information, predicting the risk level of the region and is advantageous for the related department to release early warning information. Compared with traditional surveying and mapping methods, uav surveying and mapping technology can make up for the defects of traditional surveying, especially in areas with complex terrain conditions and few people. At the same time, it can quickly provide first-hand information on disaster prevention and mitigation. In addition, uav data results can be quickly published online through mature technologies, facilitating application sharing and centralized management, shown in Fig. 3.

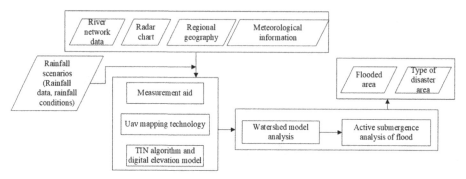

Fig. 3. Flowchart of disaster warning subsystem

Disaster Communication Subsystem. The subsystem is equipped with the Beidou Navigation satellite system and adopts the regional short message service of the Beidou-3. The RNSS signal of Beidou Navigation Constellation is used for user position calculation and single timing, and two-way message communication between end users and rescue personnel is realized through two-way data transmission capability of a CEO satellite to speed up rescue speed. Communication principle is shown in Fig. 4.

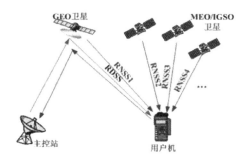

Fig. 4. RNSS + message communication system

Disaster Reporting Subsystem. This subsystem adopts "4 + 3" mode, namely four levels, three levels. The four levels are express, initial, continued, nuclear report; The three levels are county, city and province. Disaster occurred in the early and the masses life affected by a certain, certain disaster emergency management department at the county level status quote at the beginning of city emergency management departments for (when disaster situation particularly urgent, will use the letters directly report to emergency management department at the provincial level), the municipal emergency administration report summary data to the provincial department of emergency management. In the course of a disaster, local administrative departments at all levels shall report it once every seven days. After the disaster is basically stabilized, provincial emergency management departments will check it level by level and form verification reports shown in Fig. 5.

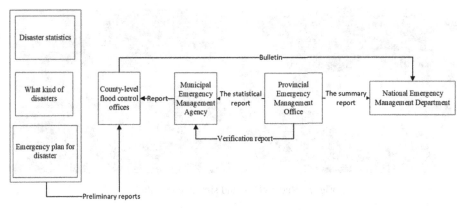

Fig. 5. Structure of "4 + 3" mode

Disaster Relief Command Subsystem. The subsystem includes establishing command center, disaster risk level and making plan. According to the data results summarized by the disaster reporting subsystem, the disaster risk level is assessed and the corresponding plan is prepared, and the number of personnel and materials is counted. One of the most important is the establishment of a command center, where rescue workers can adjust the plan according to the local actual information after arriving at the disaster area, so as to issue the most timely and accurate rescue and disaster instruction, described in Fig. 6.

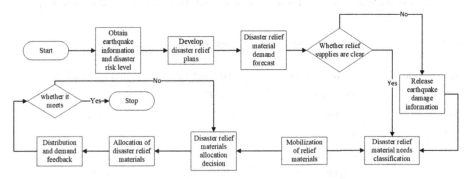

Fig. 6. Flow chart of disaster relief command subsystem

Reservoir Condition Monitoring. Different from the traditional emergency plan, the subsystem uses three-dimensional electronic sand table and LocaSpace SDK technology to divide the flood level into four risk levels, blue, yellow, orange and red, according to the relevant rainfall monitoring results, and monitor the data of rain situation monitoring points, such as reservoirs and DAMS. The system conducts dynamic simulation and analysis of precipitation according to the monitored rainfall data, builds 3D models in key areas, marks risk points, etc., and prepares corresponding plans as shown in Fig. 7 to reduce the risks of secondary disasters and associated disasters.

Fig. 7. Radar cloud and simulated rainfall

Post-disaster Summary Subsystem. The subsystem analyzes and calculates the disaster information and obtains the experience and lessons of relevant disasters. Through WheatA database export related to disasters and related factors of the geographic data analysis, analyzed the relevant factors on the prone to the influence degree of disaster area, mainly analyzes the cause of flood, find out disaster conditions, convenient for the prevention and summing up experience, on the basis of the layout related reconstruction points, reduce the difficulties caused by geographical environment in the course of reconstruction.

Factors affecting the occurrence of flood disasters are derived through WheatA, and the following data are listed in Table 5 includes factors affecting flood disasters.

Table 5. Statistics of Shanxi province in 2021

Year/Month	Surface pressure	Air temperature	Precipitation
2021.1	905/hPa	−5 °C	4 mm
2021.2	903/hPa	2 °C	6 mm
2021.3	902/hPa	6 °C	30 mm
2021.4	902/hPa	10 °C	10 mm
2021.5	895/hPa	17 °C	30 mm
2021.6	894/hPa	22 °C	45 mm
2021.7	893/hPa	24 °C	75 mm
2021.8	896/hPa	22 °C	72 mm
2021.9	900/hPa	19 °C	106 mm

Information Sharing Subsystem. This subsystem uses Oracle database to establish disaster information table (including disaster analysis table, disaster statistics table), personnel statistics table (including disaster population statistics table, dead or missing people statistics table, rescue and resettlement population statistics table) and disaster situation statistics table. Set up databases of personnel information and disaster situation, and share disaster data with the help of Internet platforms, shown in Fig. 8.

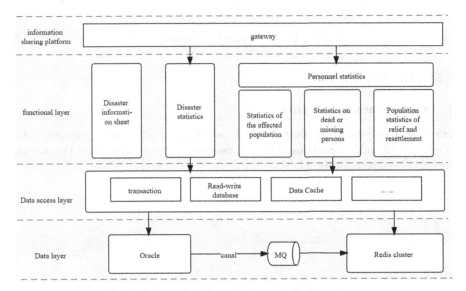

Fig. 8. Information sharing subsystem design diagram

Operations

The subsystem mainly provides users with access to the system, but also provides different types of user visual information. When there is no problem with the user's account information, the system will provide the corresponding verification code for verification, so as to prevent some illegal users from logging in and protect the information shown in Fig. 9 and 10.

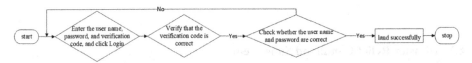

Fig. 9. System login flow chart

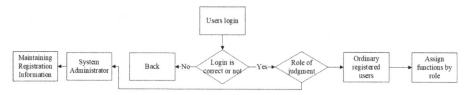

Fig. 10. User login permission management

4   System Application

4.1   Disaster Warning Subsystem

The system mainly calculates the possible disaster risk according to the real-time statistics of rain and water conditions at ground monitoring points, and issues the warning information in advance (Fig. 11).

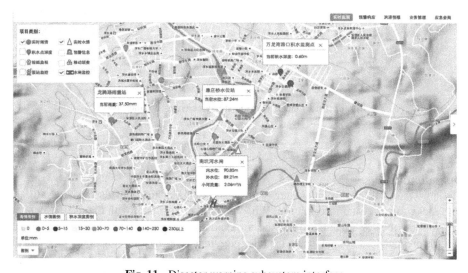

Fig. 11. Disaster warning subsystem interface

4.2   Disaster Relief Command Subsystem

According to the geographical location information of the affected area, the system quickly prepares relevant rescue plans and rescue routes, and uploads disaster logs to the system background for record (Fig. 12).

4.3   Reservoir Condition Detection Subsystem

The system is mainly for real-time monitoring of regional reservoir water level, reservoir water pressure, dam pressure and other sensitive elements. When abnormal phenomenon

Fig. 12. Disaster relief command subsystem interface

exceeds the standard, the module will inform the corresponding responsible personnel in the first time through SMS, web page and other ways, and make flood prevention plan in advance (Fig. 13).

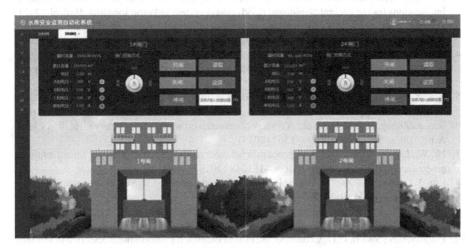

Fig. 13. Reservoir condition detection subsystem interface

5 Discussions

By comparing four kinds of satellite navigation systems, BDS Beidou navigation satellite system (with high security and independently developed in China) is finally determined as the carrying platform of flood control and disaster reduction system. The system can collect and notify disaster information through Beidou short message service. DEM is used to build a digital elevation model to fit the geographical conditions of the disaster area and calculate the risk level to issue risk warning. Through the "4 + 3" mode to achieve disaster information statistics and report to the emergency management department, the fastest release of rescue information; Three-dimensional electronic sand table and LocaSpace SDK technology are used to simulate reservoir conditions and monitor reservoir conditions in real time to reduce the risk of secondary disasters. Build a database to achieve disaster information collection.

However, the system still has shortcomings: for example, it still cannot break the barrier of "information island", cannot connect the data platform of various functional departments, and the acquisition of data information lacks timeliness and sharing, etc. In the Logistic regression analysis, only 5000 groups of data were randomly selected as analysis samples, resulting in imperfect data samples and slight differences in fitting accuracy between regression equation and actual equation. In future studies, repeated sampling and calculation are needed to remedy this deficiency.

References

1. Obi, R., Nwachukwu, M.U., Okeke, D.C., et al.: Indigenous flood control and management knowledge and flood disaster risk reduction in Nigeria's coastal communities: an empirical analysis. Int. J. Disaster Risk Reduct. **55**(2), 102079 (2021)
2. Abdelal, Q., Al-Rawabdeh, A., Qudah, K.A., et al.: Hydrological assessment and management implications for the ancient Nabataean flood control system in Petra, Jordan. J. Hydrol. (2021)
3. Duan, G., He, Q., Liu, W.: Development status and trends of foreign navigation systems. Electr. Technol. Softw. Eng. **19**, 29–31 (2021)
4. Ruiz, F.J., Ripoll, L., Hidalgo, M., et al.: Development and application of meteorological disaster monitoring and early warning platform for characteristic agriculture in Huzhou city based on GIS. Talanta (2019)
5. Xia, X., Min, J., Shen, F., et al.: Aerosol data assimilation using data from Fengyun-4A, a next-generation geostationary meteorological satellite. Atmos. Environ. **237**, 117695 (2020)
6. Sun, J., Zhang, Y., Liu, R., et al.: A review of research on warm-sector heavy rainfall in China. Adv. Atmos. Sci. **36**(12), 1299–1307 (2019)
7. Ni, W., Liu, T., Chen, S.: Rapid generation of emergency response plans for unconventional emergencies. IEEE Access **8**, 181036–181048 (2020)
8. Liu, Y., Zhang, H., Xu, Y., Wang, D.: The status and progress of global navigation satellite system. J. Navig. Position. **7**(01), 18-21+27 (2019)
9. Chen, Q., Yi, J.: Analysis of four global satellite navigation systems. J. Navig. Position. **8**(03), 115–120 (2020)
10. Fogarty, E.S., Swain, D.L., Cronin, G.M., et al.: Potential for autonomous detection of lambing using global navigation satellite system technology. Anim. Prod. Sci. **60**, 1217–1226 (2020)
11. Yu, J., Tan, K., Zhang, C., et al.: Present-day crustal movement of the Chinese mainland based on Global Navigation Satellite System data from 1998 to 2018. Adv. Space Res. **63**(2), 840–856 (2019)
12. Zhijian, L., Xiaojiang, et al.: instantaneous ambiguity resolution in global-navigation-satellite-system-based attitude determination applications: a multivariate constrained approach. IEEE Wirel. Commun. **2**, 6 (2019)
13. Zhu, Y., Tan, S., Zhang, Q., et al.: Accuracy evaluation of the latest BDGIM for BDS-3 satellites. Adv. Space Res. **64**(6), 1217–1224 (2019)
14. Devices P. BeiDou navigation satellite system. Antibiotics (2014)

Short Paper Track

Demystifying Metaverse as a New Paradigm of Enterprise Digitization

Yiyang Bian[1], Jiewu Leng[1,2], and J. Leon Zhao[3(✉)]

[1] City University of Hong Kong Shenzhen Research Institute, Shenzhen, China
[2] State Key Laboratory of Precision Electronic Manufacturing Technology and Equipment,
Guangdong University of Technology, Guangzhou, China
[3] School of Business and Management, Chinese University of Hong Kong (Shenzhen),
Shenzhen, China
leonzhao@cuhk.edu.cn

Abstract. As Facebook takes on metaverse as its new business infrastructure and company image, this new development blows a swirling wind through the IT industry. However, metaverse as a new enterprise paradigm is still quite unfamiliar to many technical people, letting along most business managers. In this paper, we give an overview of metaverse as a potential business platform and propose a framework for enterprise digitization. In particular, we clarify several critical concepts such as blockchainization, gamification, tokenization, and virtualization in relation to four Ps of marketing mix: People, Place, Product, and Process; although as many as ten Ps have been proposed, four of them are sufficient for our purposes.

Keywords: Metaverse · Enterprise digitization · Marketing mix · 4Ps

1 Introduction

Metaverse is at the forefront of the next evolution of the IT industry. At Connect 2021, the former Facebook CEO Mark Zuckerberg introduced Meta, which brings Facebook into the metaverse area[1]. The new schema that Meta plans to build will help users to communicate, collaborate, contact, and act in various business contexts. Prior to this, Microsoft Mesh was unveiled during the Microsoft's Ignite event in 2021, which is a new mixed reality platform that helps users in different geographical locations to join shared and collaborative holographic experiences and offers virtual reality meetings with enterprise-grade security. An increasing number of reputed IT companies and platforms are beginning to position themselves in the sphere of the metaverse, attempting to lead the changes that the metaverse will bring to new digital businesses.

Metaverse is an immersive virtual world where people interact as avatars with other individuals and organizations, using the metaphor of the real world but without its

---

[1] https://about.fb.com/news/2021/10/facebook-company-is-now-meta/.

© Springer Nature Switzerland AG 2022
J. Wei and L.-J. Zhang (Eds.): BigData 2021, LNCS 12988, pp. 109–119, 2022.
https://doi.org/10.1007/978-3-030-96282-1_8

physical limitations [4]. Our focus on metaverse is in its use by users and organizations, who might be physically and organizationally dispersed and rely on collaboration technologies to conduct business activities.

Metaverse has been evolving and is poised to become a new digital platform for commercial interactions. The revolutionary nature of Metaverse is likely to give rise to a range of new technological advancements in a long line of IT technologies such as artificial intelligence (AI), virtual reality (VR), augmented reality (AR), mixed reality (MR), extended reality (XR), brain computer interface (BCI), internet of things (IoT). This evolution can provide enterprises with a novel approach to collaborate with patterners and overcome existing barriers in enterprise digitalization.

Digitalization has been recognized as one of the major trends that change business in the future. According to P Parviainen et al. [10], we define enterprise digitalization in this study as changes in ways of tasks, roles, and business processes caused by the adoption of digital technologies in an organization, or in the operating environment of the organization. Today, we are experiencing another wave of changes, where the converging SMAC technologies (social, mobile, analytics, and cloud computing) are combining with increasing gigantic storage capacity and super computing power to enable new business processes and models that were unimaginable previously. To achieve this, metaverse might provide a comprehensive scenario for supporting organizations in their digital transformation and will change the business environment where enterprises operate today.

It is believed that the techniques and applications under metaverse have not been explored adequately. Meanwhile, there are still many unsolved problems for enterprises in landing on metaverse; for instance, how enterprises can get permission to collect esoteric biometric data from users, such as physiological responses, brainwave patterns for providing services? Metaverse is a brand new enterprise paradigm for many technical people as well as most business managers. Thus, our study will give an overview of metaverse development for enterprise digitization. We explore the mechanisms of digitization in the metaverse environment with several enterprise dimensions and outline a conceptual framework of metaverse-based enterprise digitization and develop a set of potential research directions in this new effort.

The remainder of this research is organized as follows. In the next section, an overview of metaverse as a new business environment is given. Then, four mechanisms of the metaverse are proposed in the context of business digitization. After that, the framework of the metaverse-based enterprise digitization is explained. The following section describes the potential research directions in this study. Finally, we conclude with a summary of the theoretical and practical implications of this study.

2 Overview of Metaverse as a New Business Environment

We have witnessed the transformation of the business environment with information technology such as electronic commerce in the past several decades. However, as valuable as online conferencing, project management, and business collaboration are, the existing tools do not fully replicate the in-office team environment. Hence, metaverse opens the door to future digital collaboration, where users can work in a virtual world that resembles the real world more and more. With metaverse, a person will be able to teleport instantly as a hologram to be at the office without a commute. Enterprise personnel will be able to spend more time on what matters to their organization, cut down time in traffic, and reduce carbon footprint.

The implementation of apps and technologies in metaverse will accelerate the progress of enterprise digitization in ways that are not possible before. Metaverse will help enterprise to expand their offline business activities into the virtual world. It will let international business collaborators share immersive experiences when they cannot be together and do collaboration activities that could not be done in the physical world without long-distance traveling.

For instance, the newly established Meta is working to bring innovative metaverse features to real-life by providing services in terms of social connection, entertainment, gaming, education, commerce, etc. Meta is now developing a high-end VR headset codenamed the project Cambria. The project Cambria will construct the capability to represent objects in the physical world with a sense of depth and perspective with the help of new sensors and algorithms.

Meanwhile, Microsoft Mesh, a mixed reality platform, intends to improve the effectiveness of collaboration and enable better team creativity. It will provide services around the areas of animated avatars, interactive meetings, app integrations, and project immersion. Mesh is one of the latest innovations in Microsoft's ecosystem for blending real and virtual worlds through mixed reality. Based on Azure platform, Mesh will also enable developers, data scientists, IT engineers, and architects to experiment with unique ideas in a flexible space.

The systematic digital solutions such as Cambria and Mesh are adding a new dimension to business collaboration. Demand for Metaverse with real-life connections in the business environment became increasingly obvious since 2020 when the pandemic pushed enterprises and teams apart. Business leaders and managers struggled to achieve the same productivity without a "face-to-face" environment. Transferring to metaverse can bring enterprises connections of virtual and real worlds in various combinations. Within a holographic space, business collaborators can interact with different people and contents in a shared and secure environment.

3 Four Dimensions of Metaverse for Business Adoption

3.1 Blockchainization

Blockchainization in this study refers to applying blockchain-based techniques in business collaboration, production, and services [12]. The blockchain technology is foreseen as the core backbone of future organizational IT infrastructure by enhancing its security, data management and process automation. A blockchain network is essentially a trustless, peer-to-peer and continuously growing database/ledger of records that have been applied among the enterprises [13]. It enables business applications and systems to operate in a fully decentralized environment without the need for any third party or trust authority.

To unlock the tremendous potential of blockchainization, several challenges will need to be addressed before this process becomes economically and legally viable in practice such as blockchains governance, data privacy, and validity of smart contracts [5]. Meanwhile, how to combine technical details and business needs in Blockchainization is also essential. The Metaverse environment will help organizations to integrate their business models and Blockchain-enabled solutions and incentive mechanisms [7].

3.2 Gamification

Enterprise gamification is one of the major human-computer interface trends of the 21st century [11]. It has significant benefits within the company in terms of (1) improving employees' work motivation; (2) better effectiveness and transparency in information transforming; (3) intensifying management objectives; (4) strengthening cooperation subject; (5) engaging better working environment and learning experiences.

Gamification is an effective way to engage and motivate younger workers and collaborators by using game-like user interfaces and applications. For instance, to encourage engagement with products, Oracle provides games such as Oracle Vanquisher, Oracle Storage Master, and Oracle x86 Grand Prix. Thus, with the development of the metaverse, employers are able to provide attractive enterprise gamification solutions for corporate innovation, motivate their employees, and manage business processes.

3.3 Tokenization in Economics

In enterprise management, people constantly hear about digital assets and tokenization. How to transfer a physical asset to a tokenized asset is still under exploring by enterprises and institutes. Currently, digital assets are usually in the forms of cryptocurrency, security token, and utility token [6]. With the development of Blockchain and IT techniques, individuals and enterprises are obtaining various digital assets. Digital assets have enabled our globalized society with the ability to efficiently transfer value.

A token is usually referring to a digital unit of a cryptocurrency that is used as a specific asset or to represent a particular use on the blockchain. Tokens have had multiple use cases which attract researchers and practitioners to explore this area. In metaverse, Tokenomics will help users to understand and evaluate how much an asset might be worth in the virtual world. Meanwhile, with the development of cryptocurrency and the NFT market, Tokenization will empower enterprises for transferring their assets/products/services independently and efficiently in a metaverse context.

3.4 Virtualization

Virtualization is key in both metaverse and enterprise digitalization. It combines the real world with the digital world, which perceives and interacts with each other in a metaverse. The techniques such as virtual reality (VR), augmented reality (AR), mixed reality (MR), extended reality (XR), brain computer interface (BCI), distributed storage (IPFS) will be used to support virtualization processes created in a virtual world.

Virtualization technology has been successfully adopted in deploying high-performance and scalable infrastructure for famous projects and applications such as Hadoop and Spark [3]. However, as virtualization is also considered a high resource-consuming technique as it usually requires a lot of storage and running space, more and more solutions are being studied and explored [1]. Within the world of the metaverse, virtualization can provide solutions for most components in the business environment such as implementing data centers with networking and collaboration at different levels. With virtualized service approach, enterprises will be able to deliver to countless customers the highest quality products and services anywhere and anytime in a metaverse.

4 Four Ps of Marketing Mix

The marketing mix is a well-known concept in academia and industries over the years that help realize marketing goals by orchestrating a hybrid decision-making behavior of mixed elements. The marketing mix framework is practical for classifying business processes and therefore it can be used to explore the impact of metaverse. This paper adopts a 4Ps framework of marketing mix and offers fresh insights into how the enterprise digitization can be configured from the perspectives of people, product, place, and process.

4.1 People

People in the enterprise are involved in the interactions with each other during the product design, process planning, system configuration, product manufacturing, quality control, delivery, management, and enterprise organization. It may include designers, planners, workers, researchers, engineers, officers, managers, scientists, and a wide array of other human resources trained to provide manufacturing or product services. Enterprises need to effectively manage their personnel to monitor, maintain, and/or upgrade their manufacturing or product service quality concerning attitude, competency, and professionalism, all of which are equally important for customer experience and satisfaction.

The new generation of IT technologies gives people the power to connect and express themselves more diversely and naturally. The metaverse blueprint will unlock a massively larger creative economy than the one constrained by today's platforms and their policies. However, a metaverse will not break away from their existing social media networks.

4.2 Product

Product encapsulates the service solutions offered and sold by enterprises. These products consist of physical goods and intangible services that promote and maintain the quality and performance of the product for obtaining added value. Enterprises must deliver and communicate the minimum level of the expected performance of product service quality.

In the metaverse blueprint, the product becomes more immersive and will be touched at any time and in any manner. A shift happens towards the embedding of online service with new social experience, flexibility, co-manufacturing, and creative opportunities. Metaverse could provide more individualized products with up-to-date styles, variety, and new releases that are not available in the real world. Metaverse is a platform for creativity that enables the display of full details and information about a product, which is critical to user experiences.

4.3 Place

The place is an important element of the enterprise digitization. "Place" could be defined from three perspectives: 1) place as a spatial concept, 2) place as an anthropological concept, and 3) place as a contemporary concept [2]. The facility in the place is an outlet for social interaction as well as the use of service to optimize the experiences. "Shared place" is an extended concept in cyberspace, wherein interaction can happen. The place is socially determined and thus is not a determinant property of virtual worlds [8].

The modeling capability and defining quality of the metaverse will make people feel a presence and social interaction with another person or do almost any work with partners in any place you can imagine in a single physical location. The existed monitors, work setup, control panels, and more physical things assembled in factories could just be holograms designed by creators in the metaverse blueprint. The engineers will be able to teleport remote production sites instantly as a hologram to be at the office without a commute, at a concert with partners. The workers could move across different sites and rapidly obtain fully-immersed experiences via augmented reality glasses while staying present in the physical world. Therefore, the workers can spend more time on what matters to the production, cut-down time in traffic and screens, and reduce the carbon footprint. Metaverse makes it possible to be a participant regardless of the place of residence, location, and other space constraints. This positive fact undoubtedly contributed to the socialization of enterprise under digitization.

4.4 Process

Resources alone do not account for an enterprise's competitiveness. The ability of an enterprise to orchestrate its resources and optimize performance is embodied in business and managerial processes [9]. Process in an enterprise describes the operating and tracking procedures and systems by which enterprises deliver product and manufacturing services. Timely optimizing the processes in the enterprise can help cut-down costs, as well as improve monetary profit, non-monetary brand reputation, service satisfaction, and recommendation returns. The process is a key management element to deliver value and achieve the sustainability goals of the enterprise. The production manufacturing is now moving towards servitization, and under the metaverse blueprint, the product service, as well as manufacturing service, will be embodied with completely new experiences.

Generally, the 4Ps of enterprise digitization offer decision-makers a set of decision variables to (1) strategically present the promise of their product and manufacturing services to potential partners and customers; (2) satisfy the individualized requirements of these users; (3) provide guidelines for making desired behavioral responses from these users; and (4) accomplish established development goals.

Table 1. Innovation cases under metaverse-based enterprise digitization

	Blockchainization	Gamification	Tokenization	Virtualization
People	Enabling machine trust to substitute for institutional trust in metaverse-based transactions such as credit rating and information encryption	Innovative business interactions and engagements in metaverse enriched by gamified designing such as point systems and badges	Tokenized features enhancing business activities and collaborations in service digitization for metaverse tasks	Virtualization connects the real world with cyberspace for providing users with new functional features such as teleport
Product	Trusted product information sharing by increasing the data transparency and traceability such as cold chain logistics and risk management	Design of product features among end-users via simulation such as digital try-on and custom-tailored in virtual marketing	Sharing of product usage via various payment methods by the realization of digital asset and the development of token economics	Improved vitality and compatibility in product creation, feature design, and usage experience through virtualized approaches
Place	Prevention of data frauds from the metaverse to real world by deploying a decentralized platform for data interaction and manipulation	Simulated games in the virtual world relating to real places and providing location-based services/interactions such as Pokémon GO	Tokenomics transferring the location of stakeholders from one place to distributed places for better ecosystem sustainability	The shared place in a metaverse mapping between the virtual site and real locations anywhere from the world
Process	Smart contracts connect validated business tasks, ensuring the data management consistency and process fluency	A business process with gamified simulation involving more user-generated contexts and providing more genres and segments	Security control in metaverse by tokenized processes enhancing decentralization of authority and eliminating market entry barriers	Higher-level immersive processes of collaboration, and interaction such as horizon workrooms and immersive shopping

Note: In the conceptual analysis of innovation cases in Table 1, we assume a business context where merchandises are designed, produced, marketed, transported, sold, and serviced with the application of metaverse technology and systems.

5 A Framework of Enterprise Digitization in a Metaverse Environment

Metaverse is a virtual world built on cyberspace with growing maturity of various virtual reality technologies. A virtual world is both mapped to and independent from the real world. Meanwhile, the metaverse is not simply a virtual space but encompasses networks, hardware terminals, and users in a sustainable, wide-reaching virtual reality system that includes digital replicas of the real world as well as virtual world creations. A metaverse is to be created and built by multiple creators and developers, leading to new experiences and digital items that are interoperable.

Based on analysis of four mechanisms of metaverse for business adoption and 4 Ps of enterprise digitization, this paper proposes a conceptual matrix of metaverse-based enterprise digitization as shown in Table 1. The logic of this conceptual matrix can be understood from the blockchainization, gamification, tokenization, and virtualization perspectives. Table 1 gives some potential innovation cases in metaverse-based enterprise digitization.

Blockchainization, as the key factor in the metaverse, enables decentralized peer-to-peer interaction. The decentralized blockchain network could provide unique identification for each created data via generated hash codes.

- *People.* Blockchainization in metaverse will empower people with machine trust and encrypted information sharing. The trust of interactions among entities in the metaverse will be realized by a novel credit rating system via blockchain technology naturally.
- *Product.* Blockchain will improve an enterprise's product information sharing by increasing the transparency for customers in the metaverse. The increased trust in product information means that customers will be able to know products' traceable information for assessing the potential risk and delays in shipping.
- *Place.* A shared place in a metaverse for enterprises will be boundary-breaking and formulated as a virtual ecosystem. Blockchainization can also prevent data frauds from being metaverse to real locations by deploying a decentralized network for manipulating and transferring crucial business data.
- *Process.* The process of metaverse cannot be paused. To assure data is not lost in the business process, the metaverse is supposed to enable the persistence of all data [8]. The process in a metaverse for enterprises will be characterized by smart contracts designing and being validated in blockchain networks.

Gamification is an advanced paradigm of enterprise competition in the metaverse blueprint. Many cases of metaverses started as multiplayer online role-playing games and rapidly evolved into alternative realities.

- *People.* A new social interaction context has emerged with the characteristics of gamification. For instance, earning credit-based user "points" and "badges" in daily activities. The gamification will empower people in metaverse with new features of social pleasure, advanced networking capability, and game-style engagement.

- *Product.* Products in metaverse will be enhanced with more creativity, productivity, and diversity among end-users via simulation. For instance, the services of digital try-on and custom-tailored are supported by several online platforms. Users could customize and provide designing features via gamified simulation.
- *Place.* The shared places in a metaverse for enterprises will be a mix of online/offline communication. With the development of AR and MR, more and more online simulated games providing location-based interactions for players such as *Pokémon GO*. Predictably, the virtual places for enterprise collaboration will be also more differentiable and diverting.
- *Process.* The design of the business process with gamification will involve more user-generated contexts. The business process will provide more genres and business segments by gamified simulation in scenario-based workflow management and job training.

Tokenization is a critical enabling method for promoting the participation of socialized resources and accelerating crowd intelligence in metaverse. Tokenization will help enterprises for providing their products and services efficiently in a metaverse context.

- *People.* Tokenization will accelerate service digitization for users' metaverse tasks and business processes. It will empower the people in metaverse with new tokenized features and more possibilities of achieving win-win cooperation for individual users and enterprises.
- *Product.* Product usage in metaverse could be shared via various payment methods such as micropayments, which will enhance digital asset realization. The product creation in metaverse will be also enhanced by more self-motivated designers via the development of token economics.
- *Place.* The shared place in a metaverse for enterprise will be more distributed. Tokenomics will transfer the location of stakeholders from one place to distributed places and make the ecosystem in virtual world more defensible and sustainable. For market orientation, it will evolve from the conventional "value of brands" to "value of locations" in the metaverse.
- *Process.* The security control in metaverse could be also improved by tokenized processes. Process in metaverse will be characterized by high-level decentralization of authority, low market entry barriers, and high sustainability.

Virtualization is the bridge for connecting the real-world experience with the dedicated scenario design in cyberspace. Virtualization technologies can be used for the simulation of face-to-face interaction, knowledge sharing, collaboration across distributed places, terminals, and contexts.

- *People.* Virtualization enables people in the metaverse with new features teleport. As such, virtualization will make an incredibly commercial place for different participates within a business to design more vivid and compatible artifacts.
- *Product.* The creation and use of the product in metaverse will be virtualized. The vitality and compatibility of products in metaverse will be considered in creating

new products with the metaverse environment. Meanwhile, the feature designing in a virtual world will also improve offline production.

- *Place.* Enterprises can reach this shared place in a metaverse anywhere from the world. And there exist specific mapping rules for all the enterprises to explore between the metaverse and their real locations.
- *Process.* Process in a metaverse for enterprises will be enhanced by higher-level immersive collaboration and low friction. For employees, *Horizon Workrooms* created by Meta provides a way for office workers to connect using virtual reality. For consumers, the immersive experiences of virtual shopping will become extremely self-directed, high-touch, and consultable whenever and wherever they want.

In terms of the challenges of implementing a metaverse blueprint, it is not difficult to build a virtual converged reality space in metaverse based on the current IT technology, but the difficulty is on how to replicate or map to the real world diligently. The realization of metaverse cannot be completed by one company alone, because the industrial ecology, hardware performance, and system optimization are extremely complex. It will involve unexpected problems such as how to satisfy the extremely-high network bandwidth, how to establish the supply chain ecology, and how to design the operating rules and behave ethics of the metaverse.

6 Concluding Remarks

Metaverse has a positive effect on transformation processes in the activities of enterprise digitalization. In particular, metaverse-based business and technology investments will result in significant improvements in activity efficiency of both the individual employee and the activity of the enterprise as a whole in terms of the four Ps. This is an incredibly exciting time for realizing enterprise digitization in metaverse by advancing multiple research areas such as blockchain, AI, VR, BCI, etc. Many of these research areas are currently being explored in parallel to transform how we think of business.

This study gives an overview of metaverse as a potential business platform and proposes a framework for enterprise digitization. However, there are some risks and disadvantages posed by metaverse in businesses digital transformation process. For instance, the increase in the number of unemployed; high dependency on the leading platform in the metaverse, and potential costs in transferring and protecting their corporate data in a virtual world. In future research, more work is needed to prototype and test metaverse-based solutions for enterprise digitalization and business transformation.

For research scholars, we direct the attention of IS research toward the metaverse-driven business innovation, which is characterized by recursive interactions between enterprises and the metaverse technology. Rooted in the four Ps of the marketing mix, we offer a framework of enterprise digitization in a metaverse environment that underscores the need for conceptualizing academic problems in the existing areas around metaverse. For IT and business practitioners, we hope the conceptual and technical viewpoints highlighted in this paper will help with determining what is possible when the metaverse is applied to enterprise activities and with considering how these business indicators and metaverse features lead to changes in enterprise digitalization.

Acknowledgement. This work is partially supported by NSFC Grants No. 72031001, 72104104, 71932002, Shenzhen S&T Innovation Fund No. JCYJ20170818100156260, and Hong Kong ITF Fund GHP/142/18GD.

References

1. Aguiar, A., Hessel, F.: Embedded systems' virtualization: the next challenge? In: Proceedings of 2010 21st IEEE International Symposium on Rapid System Protyping, pp. 1–7. IEEE (2010)
2. Anderson, E.D.: The maintenance of masculinity among the stakeholders of sport. Sport Manag. Rev. **12**, 3–14 (2009)
3. Bhimani., J, Yang, Z., Leeser, M., et al.: Accelerating big data applications using lightweight virtualization framework on enterprise cloud. In: 2017 IEEE High Performance Extreme Computing Conference (HPEC), pp. 1–7. IEEE (2017)
4. Davis, A., Murphy, J.D., Owens, D., et al.: Avatars, people, and virtual worlds: foundations for research in metaverses. J. Assoc. Inf. Syst. **10**, 90 (2009)
5. Hamida, E.B., Brousmiche, K.L., Levard, H., et al.: Blockchain for enterprise: overview, opportunities and challenges. In: The Thirteenth International Conference on Wireless and Mobile Communications (ICWMC 2017) (2017)
6. Howell, S.T., Niessner, M., Yermack, D.: Initial coin offerings: financing growth with cryptocurrency token sales. Rev. Financ. Stud. **33**, 3925–3974 (2020)
7. Lacity, M.C.: Addressing key challenges to making enterprise blockchain applications a reality. MIS Q. Executive **17**, 201–222 (2018)
8. Nevelsteen, K.J.: Virtual world, defined from a technological perspective and applied to video games, mixed reality, and the Metaverse. Comput. Anim. Virtual Worlds **29**, e1752 (2018)
9. Orji, C.I.: Digital business transformation: towards an integrated capability framework for digitization and business value generation. J. Global Bus. Technol. **15**, 47–57 (2019)
10. Parviainen, P., Tihinen, M., Kääriäinen, J., et al.: Tackling the digitalization challenge: how to benefit from digitalization in practice. Int. J. Inf. Syst. Proj. Manag. **5**, 63–77 (2017)
11. Rauch, M.: Best practices for using enterprise gamification to engage employees and customers. In: Kurosu, M. (ed.) Human-Computer Interaction. Applications and Services, vol. 8005, pp. 276–283. Springer, Cham (2013). https://doi.org/10.1007/978-3-642-39262-7_31
12. Zhao, J.L., Fan, S., Yan, J.: Overview of business innovations and research opportunities in blockchain and introduction to the special issue. Financ. Innov. **2**(1), 1–7 (2016). https://doi.org/10.1186/s40854-016-0049-2
13. Zheng, Z., Xie, S., Dai, H.-N., et al.: Blockchain challenges and opportunities: a survey. Int. J. Web Grid Serv. **14**, 352–375 (2018)

Author Index

Printed in the United States
by Baker & Taylor Publisher Services

Printed in the United States
by Baker & Taylor Publisher Services